Lecture Notes in Computer Science **16293**

The series Lecture Notes in Computer Science (LNCS), including its subseries Lecture Notes in Artificial Intelligence (LNAI) and Lecture Notes in Bioinformatics (LNBI), has established itself as a medium for the publication of new developments in computer science and information technology research, teaching, and education.

LNCS enjoys close cooperation with the computer science R & D community, the series counts many renowned academics among its volume editors and paper authors, and collaborates with prestigious societies. Its mission is to serve this international community by providing an invaluable service, mainly focused on the publication of conference and workshop proceedings and postproceedings. LNCS commenced publication in 1973.

Zhaolin Chen · Sanuwani Dayarathna ·
Kh Tohidul Islam · Himashi Peiris ·
Parisa Zakavi · Shenjun Zhong
Editors

Enhancing Ultra-Low-Field MRI with Paired High-Field MRI Comparisons for Brain Imaging

First International Challenge, ULF-EnC 2025, Held in Conjunction with MICCAI 2025
Daejeon, South Korea, September 23, 2025
Proceedings

Springer

Editors
Zhaolin Chen
Monash University
Clayton, VIC, Australia

Kh Tohidul Islam
Monash University
Clayton, VIC, Australia

Parisa Zakavi
Monash University
Clayton, VIC, Australia

Sanuwani Dayarathna
Monash University
Clayton, VIC, Australia

Himashi Peiris
Monash University
Clayton, VIC, Australia

Shenjun Zhong
Monash University
Clayton, VIC, Australia

ISSN 0302-9743 ISSN 1611-3349 (electronic)
Lecture Notes in Computer Science
ISBN 978-3-032-23343-1 ISBN 978-3-032-23344-8 (eBook)
https://doi.org/10.1007/978-3-032-23344-8

This Springer imprint is published by the registered company Springer Nature Switzerland AG
The registered company address is: Gewerbestrasse 11, 6330 Cham, Switzerland

Preface

The Ultra-Low-Field MRI Image Enhancement Challenge (ULF-EnC 2025), formally titled *Enhancing Ultra-Low-Field MRI with Paired High-Field MRI Comparisons for Brain Imaging*, was held in conjunction with MICCAI 2025 in Daejeon, South Korea, on September 23, 2025. The challenge addressed a timely and clinically significant problem: bridging the quality gap between ultra-low-field (64mT) and high-field (3T) brain MRI through algorithmic enhancement using paired imaging data.

Ultra-low-field MRI systems are increasingly deployed in point-of-care and resource-limited settings due to their portability and affordability. However, reduced spatial resolution and contrast limitations remain significant barriers to broader diagnostic adoption. ULF-EnC 2025 was designed to provide a rigorous, benchmark-driven platform for developing and evaluating enhancement algorithms that preserve anatomical fidelity while improving image quality. The challenge dataset consisted of paired 3T and 64mT brain MRI scans (T1, T2, and FLAIR) acquired under ethically approved protocols, enabling structured cross-field comparison and quantitative assessment.

The challenge was fully hosted on the Synapse platform, which supported participant registration, secure data distribution, submission management, validation, and final testing. Throughout the challenge lifecycle, an online discussion forum and public validation leaderboard were maintained to ensure open communication, transparency, and timely feedback to participants. During the validation phase alone, more than 430 validation submissions were processed. After each submission, validation scores were automatically computed and communicated to teams via email, and the leaderboard was updated accordingly to provide continuous performance tracking.

Following the validation phase and paper submission deadline, the final testing phase was opened. Participating teams submitted Docker containers to ensure standardized and reproducible evaluation. All containers were executed under controlled computational constraints, and final results were published on the testing leaderboard. This structured workflow ensured fairness, reproducibility, and consistent comparison across all methods.

The challenge attracted strong international interest, with more than 100 registered participants from across the global medical imaging and machine learning communities. A total of 21 short papers were submitted to the proceedings. All submissions underwent a double-blind peer-review process managed through the Microsoft CMT system, following standards comparable to those of MICCAI conferences. Each paper was reviewed by a minimum of three independent reviewers, and an additional meta-review was conducted by a member of the organizing committee. Final decisions were made collectively by the chairs based on the reviewers' evaluations and recommendations. Nineteen short papers were accepted for publication in this volume. There were no invited papers, and all accepted contributions were subject to the same review criteria and quality standards.

No organizing committee members submitted papers to the challenge, thereby ensuring the absence of conflicts of interest in the review and decision-making process.

Performance ranking was determined using a weighted combination of quantitative metrics assessing structural similarity, signal fidelity, and reconstruction accuracy. The three top-performing teams were formally announced during the MICCAI 2025 challenge session and recognized with monetary awards: First Prize (A$1,500), Second Prize (A$750), and Third Prize (A$500). These awards acknowledged scientific excellence, methodological rigor, and computational efficiency.

We gratefully acknowledge the support of the National Imaging Facility (NIF), an NCRIS-enabled capability, for providing infrastructure and technical expertise that made this challenge possible. We also thank Hyperfine Inc. for their support and all members of the program committee and external reviewers for their careful and constructive evaluations. Their efforts ensured the scientific rigor and quality of this proceedings volume.

We hope that the contributions collected in this LNCS volume will serve as a valuable benchmark reference for the emerging field of ultra-low-field MRI enhancement and will stimulate further research toward equitable, accessible, and clinically reliable imaging solutions worldwide.

September 2025

Zhaolin Chen
Sanuwani Dayarathna
Kh Tohidul Islam
Himashi Peiris
Parisa Zakavi
Shenjun Zhong

Organization

General Chair

Zhaolin Chen	Monash University, Australia

Program Committee Chairs

Sanuwani Dayarathna	Monash University, Australia
Kh Tohidul Islam	Monash University, Australia
Himashi Peiris	Monash University, Australia
Parisa Zakavi	Monash University, Australia
Shenjun Zhong	Monash University, Australia

Additional Reviewers

Levente Baljer
Geng Chen
Bruno de Almeida
Kaiyuan Deng
Xiangyao Deng
James Grover
Peter Hsu
Baris Imre
Seonghyuk Kim
Youngmin Kim
Jeongchan Kim
Snehil Kumar
Taehoon Lee
Jinghang Li
Yiyang Lin
Zhicheng Lu
Alfredo Lucas
Jan Nikolas Morshuis
Haowen Pang
Sung-Hong Park
Zhiqiang Shen
Jaeyun Shin
Neil Vaughan
Mengfei Wang
Sizhe Wang
Andrew Webb
Zhenyu Xiang
Ghulam Yar
Ruipeng Zhang
Felix Zimmermann

Contents

Robust and Lightweight Low-to-High Field MRI Synthesis Guided by Semantic Prior

Yiyang Lin[1], Byungjun Kim[2], Yixuan Yuan[1(✉)], and Jinglei Lv[2(✉)]

[1] The Chinese University of Hong Kong, Ma Liu Shui, Hong Kong
yxyuan@ee.cuhk.edu.hk
[2] The University of Sydney, Sydney, Australia
jinglei.lv@sydney.edu.au

Abstract. Low-field magnetic resonance imaging (MRI) has become the preferred non-invasive diagnostic modality in many medical institutions due to its low cost, compact size, portability, and high safety. However, its inherently low signal-to-noise ratio (SNR) and long acquisition time significantly hinder diagnostic efficiency and accuracy. In contrast, high-field MRI offers a substantially higher SNR, richer soft-tissue contrast, and more detailed structural information, which are beneficial for fine-grained diagnosis. Nevertheless, the high cost, bulky equipment, and limited patient compatibility restrict its widespread clinical adoption. To address these challenges and enable the rapid acquisition of high-quality images at a lower cost, we propose a method for Low-field to High-field MRI synthesis. Specifically, we design a lightweight 3D-UNet architecture capable of processing full 3D MRI volumes, thereby avoiding the spatial discontinuity issues introduced by conventional patch-by-patch processing. This design preserves semantic continuity across the entire brain, and facilitates the modeling of cross-patch interactions such as the interactions between different brain functional areas. Moreover, traditional point-to-point loss functions lack of robustness to misalignment between the input and the ground truth, which is common in MRI-based datasets and may introduce severe artifacts. To overcome this, we introduce a robust prior-based semantics matching mechanism, which leverages global-view sampling and domain knowledge to assess the semantic consistency between generated and target images without requiring voxel-wise alignment. Experimental results demonstrate the effectiveness of our approach across multiple MRI datasets, achieving a Structural Similarity Index Measure (SSIM) of 0.9112 (T1), 0.9210 (T2), and 0.8850 (FLAIR), as well as a Peak Signal-to-Noise Ratio (PSNR) of 25.54 dB (T1), 26.23 dB (T2), and 25.27 dB (FLAIR), validating the proposed method as a practical and efficient solution for enhancing low-field MRI.

Keywords: Low-to-High Field MRI Syneresis · Lightweight 3D U-Net · Robust Prior-Based Semantics Matching Mechanism

Y. Lin and B. Kim— Co-first author.

Z. Chen et al. (Eds.): ULF-EnC 2025, LNCS 16293, pp. 1–8, 2026.
https://doi.org/10.1007/978-3-032-23344-8_1

1 Introduction

Magnetic Resonance Imaging (MRI) is a cornerstone of modern medical imaging, offering non-invasive, radiation-free visualization of internal anatomical structures with exceptional soft-tissue contrast [1]. In recent years, low-field MRI systems—typically operating below 0.5 T—have garnered increasing attention in clinical and research settings due to their affordability, smaller physical footprint, portability, and enhanced safety profile, particularly for patients with implants or those requiring frequent imaging [2,3]. These advantages make low-field MRI a promising solution for resource-limited environments, point-of-care diagnostics, and mobile healthcare applications. However, the major drawback of low-field MRI lies in its inherently reduced signal-to-noise ratio (SNR) and prolonged acquisition times, which often translate to lower image quality, blurred anatomical boundaries, and limited diagnostic reliability [4]. These limitations are especially detrimental in applications requiring high spatial resolution and precise tissue delineation, such as neuroimaging and cardiovascular assessment [5,6].

In contrast, high-field MRI systems (typically 1.5T and above) offer superior SNR, enhanced tissue contrast, and finer structural detail, enabling more accurate and confident diagnoses [7]. Yet, their prohibitive cost, large infrastructural requirements, and restricted patient compatibility (e.g., claustrophobia, contraindicated implants) pose significant barriers to widespread deployment—particularly in rural hospitals, and developing regions [8].

To bridge this performance gap while retaining the accessibility of low-field devices, we propose a novel low-field to high-field MRI synthesis framework. At the core of our approach is a lightweight 3D-UNet [9] architecture designed to process entire 3D MRI volumes, thus eliminating the discontinuities and context fragmentation associated with conventional patch-by-patch methods. This holistic 3D perspective is crucial for preserving semantic continuity and capturing long-range anatomical dependencies, including interhemispheric interactions that are critical for accurate brain mapping [10–12].

Moreover, considering the imperfect alignment between low-field and high-field image pairs—often due to differences in acquisition protocols, patient motion, or scanner variability—we introduce a prior-guided, misregistration-agnostic loss function [13]. Unlike traditional voxel-wise losses that are sensitive to spatial mismatches and prone to generating artifacts, our loss formulation leverages global semantic similarity and domain-specific priors to promote perceptually coherent synthesis without requiring exact anatomical alignment [14].

Our experimental evaluations on multiple benchmark MRI datasets demonstrate that the proposed method not only achieves significant improvements in image quality, as measured by SSIM and PSNR, but also enhances the clinical interpretability of synthesized high-field-like images. These results highlight the potential of our approach to democratize high-quality MRI diagnostics, making them more accessible and practical across a range of medical contexts.

2 Method

2.1 Basic Framework of Lightweight 3D U-Net

Given a low-field 3D MRI volume as input, our goal is to generate a corresponding high-field MRI volume that preserves anatomical accuracy and fine-grained structure. To this end, we adopt a lightweight 3D U-Net architecture as our generator, consisting of an encoder-decoder structure with skip connections, as illustrated in Fig. 1. The encoder extracts multi-scale semantic and spatial features from the input, while the decoder progressively reconstructs the high-field volume. The network is supervised using a pixel-wise L2 loss (L_V) between the generated image and the ground truth high-field image to ensure voxel-level fidelity. This basic reconstruction loss enforces the network to minimize the intensity differences between the predicted and real volumes.

2.2 Robust Prior-Based Semantics Matching Mechanism

Although pixel-wise losses like L2 are widely adopted in medical image reconstruction tasks, they suffer from a critical limitation: assuming strict spatial alignment between input and target. In practice, especially in low- to high-field domain translation, small misalignments or anatomical deformations are inevitable due to variations in magnetic field strength, patient motion, or acquisition inconsistencies. As a result, L2 loss (L_V) may lack of robustness to the misalignments and can lead to undesirable artifacts such as ghosting, blurring, or anatomical hallucinations, which may impair clinical interpretation.

To address this, we propose a robust prior-based semantics matching mechanism, as shown in the blue box of Fig. 1. This mechanism contains a semantic-aware loss function (L_S) that compares the semantic similarity of generated and ground truth images, rather than enforcing pixel-wise correspondence. This design is inspired by the observation that human perception and clinical evaluation rely more on structural and anatomical consistency than pixel-perfect alignment. Our semantic loss extracts features from intermediate layers of a pretrained 2D AlexNet [15,16], representing each image in a high-level semantic space.

However, since AlexNet is a 2D network and 3D feature extraction is memory-intensive, we propose a novel **Full-view sampling (FS)** strategy. Specifically, we extract diagnostically most informative slices—typically located near the center—along each of the three orthogonal anatomical planes (axial, sagittal, and coronal) from both the generated and ground-truth volumes. These representative slices are then passed through the pretrained 2D AlexNet to compute the semantic loss. This approach ensures comprehensive semantic coverage while maintaining computational efficiency.

2.3 Loss Function Formulation

Our total loss function combines the voxel-wise reconstruction loss and the semantic-aware loss:

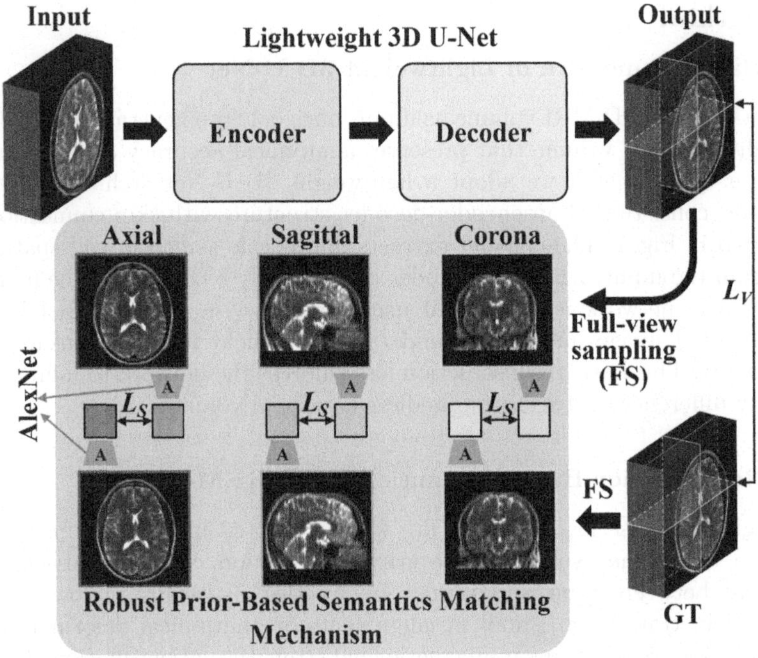

Fig. 1. Overview of our method. A 3D U-Net takes low-field MRI as input and reconstructs high-field MRI. We enforce supervision using both voxel-wise L2 loss and our proposed semantic-aware loss computed on central slices in three orthogonal views.

$$\mathcal{L}_{\text{total}} = \mathcal{L}_V + \mathcal{L}_S. \tag{1}$$

Voxel-wise L2 Loss:

$$\mathcal{L}_V = \frac{1}{N}\sum_{i=1}^{N}\left\|\hat{V}_i - V_i\right\|_2^2, \tag{2}$$

where $\hat{V}_i$ and V_i denote the i-th voxel from the generated and ground truth volumes respectively, and N is the total number of voxels.

Semantic-Aware Loss: In this loss function, for each anatomical direction $d \in \{\text{axial}, \text{sagittal}, \text{coronal}\}$, the 2D slices $I^{(d)}$ and $\hat{I}^{(d)}$ are extracted by taking the central slice from the ground-truth high-field volume V and the predicted volume $\hat{V}$, respectively. Each slice is then passed through a pre-trained AlexNet to obtain deep features $S^{(d)} = \{S_1^{(d)}, S_2^{(d)}, S_3^{(d)}, S_4^{(d)}, S_5^{(d)}\}$, corresponding to the first five convolutional layers of the network [17]. The semantic loss is defined as:

$$\mathcal{L}_S = \sum_{d}\sum_{i=1}^{5}\left\|S_i^{(d)}\left(\hat{I}^{(d)}\right) - S_i^{(d)}\left(I^{(d)}\right)\right\|_2^2. \tag{3}$$

In summary, our method leverages both low-level voxel supervision and high-level semantic supervision to improve the anatomical fidelity and perceptual quality of generated high-field MRIs, while mitigating artifacts caused by the spatial misalignment.

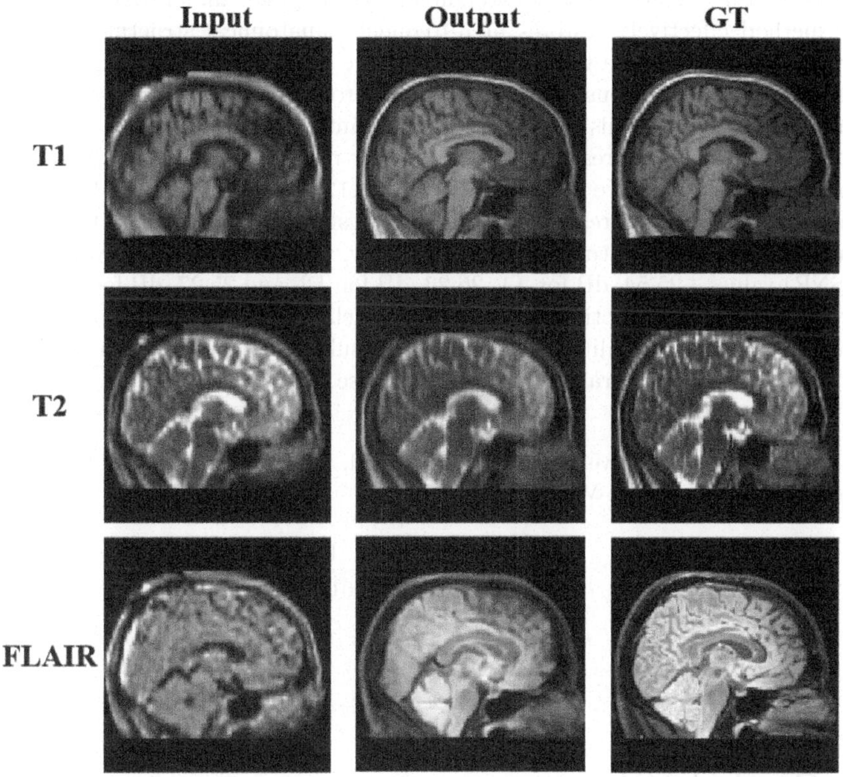

Fig. 2. The visualization results of our method, including synthesized outputs for T1-weighted, T2-weighted, and FLAIR MRI modalities.

3 Experiments

The dataset consists of 50 paired low-field and high-field MRI volumes. The challenge data was acquired using a Siemens Biograph mMR (3T) scanner and a Hyperfine Swoop (64mT) scanner. To evaluate the generalization performance of our method, we randomly selected 10 subjects as the test set and used the remaining 40 for training, following an 80%–20% split. Specifically, the test set includes the following subjects: POCEMR001, 008, 018, 023, 035, 054, 073, 088, 098, and 102. All experiments are conducted using this fixed split to ensure reproducibility [18–21]. Moreover, our model is implemented in Python (PyTorch) and runs on an NVIDIA RTX 4090 GPU. The training utilizes the Adam optimizer ($\beta_1 = 0.5$, $\beta_2 = 0.999$), with an initial learning rate of 0.0001, which linearly decays after half of the total epochs (1,000 epochs). The batch size is set to 1.

The qualitative results, as illustrated in Fig. 2, show that the synthesized high-field MR images are visually consistent with the ground truth (GT) across all three modalities—T1-weighted, T2-weighted, and FLAIR. Notably, our method effectively restores high-frequency anatomical structures, such as cortical folds and tissue boundaries, which are often blurred in the input low-field images. This demonstrates the model's strong capability in recovering fine-grained structural details and preserving semantic integrity.

Quantitatively, as presented in Table 1, our method achieves high Structural Similarity Index Measure (SSIM) values of 0.9112, 0.9210, and 0.8850 for T1, T2, and FLAIR modalities, respectively, indicating strong alignment with the ground truth in terms of structural fidelity. Similarly, the Peak Signal-to-Noise Ratio (PSNR) values—25.54, dB for T1, 26.23, dB for T2, and 25.27, dB for FLAIR—further confirm the effectiveness of our approach in generating perceptually and quantitatively high-quality images. These results validate the robustness and generalizability of our framework across diverse MRI contrasts.

Table 1. The quantitative results of our method, including results for T1-weighted, T2-weighted, and FLAIR MRI modalities.

	T1	T2	FLAIR
SSIM (↑)	0.9112	0.9210	0.8850
PSNR (↑)	25.54	26.23	25.27

4 Conclusion

In this work, we present a novel framework for enhancing low-field MRI by synthesizing corresponding high-field images using a lightweight 3D U-Net architecture. Our approach addresses the limitations of traditional voxel-wise losses by introducing a semantic-aware, misregistration-agnostic loss function that leverages domain knowledge and multi-view information. Specifically, we propose a full-view sampling strategy that extracts central slices across three anatomical planes and compares their high-level semantic features using a pretrained 2D AlexNet. This enables robust and perceptually consistent supervision without requiring explicit voxel-to-voxel alignment.

Experimental results on multiple benchmark datasets demonstrate the effectiveness of our method in improving both quantitative image quality metrics (e.g., SSIM and PSNR) and qualitative anatomical fidelity. Moreover, the proposed semantic-aware loss significantly reduces spatial artifacts and enhances the clinical interpretability of the synthesized images. These findings validate the potential of our method as a practical and efficient solution for advancing low-field MRI, particularly in resource-constrained settings where high-field systems are not readily available. Future work may explore extending this framework to other imaging modalities and incorporating self-supervised or unsupervised learning paradigms to further reduce dependency on paired data.

References

1. Najjar, R.: Clinical applications, safety profiles, and future developments of contrast agents in modern radiology: a comprehensive review. Iradiology **2**(5), 430–468 (2024)
2. Marques, J.P., Simonis, F.F.J., Webb, A.G.: Low-field mri: an mr physics perspective. J. Magnetic Resonance Imaging **49**(6), 1528–1542 (2019)
3. Sarracanie, M., Salameh, N.: Low-field mri: how low can we go? a fresh view on an old debate. Front. Phys. **8**, 172 (2020)
4. Ayde, R., Vornehm, M., Zhao, Y., Knoll, F., Wu, E.X., Sarracanie, M.: Mri at low field: A review of software solutions for improving snr. NMR Biomed. **38**(1), e5268 (2025)
5. Yang, S., Jiarui, W., Liu, J., Li, X., Zhang, Q., Pan, M., Gan, Y., Chen, Z., Zhang, S.: Exploring sparse visual prompt for domain adaptive dense prediction. In: Proceedings of the AAAI Conference on Artificial Intelligence 38, pp. 16334–16342 (2024)
6. Yang, S., Li, J., Lai, X., Yu, B., Zhao, H., Jia, J.: Visionthink: Smart and efficient vision language model via reinforcement learning. arXiv preprint arXiv:2507.13348 (2025)
7. Ladd, M.E., et al.: Pros and cons of ultra-high-field mri/mrs for human application. Progress Nuclear Magnetic Resonance Spectroscopy **109**, 1–50 (2018)
8. Rechel, B., et al.: Hospitals in rural or remote areas: An exploratory review of policies in 8 high-income countries. Health Policy **120**(7), 758–769 (2016)
9. Xiao, Z., Liu, B., Geng, L., Zhang, F., Liu, Y.: Segmentation of lung nodules using improved 3d-unet neural network. Symmetry **12**(11), 1787 (2020)
10. Sanjid, K.S., Hossain, M.T., Shahariar Junayed, M.S., Uddin, M.M.: Integrating mamba sequence model and hierarchical upsampling network for accurate semantic segmentation of multiple sclerosis legion. arXiv preprint arXiv:2403.17432 (2024)
11. Liu, S., et al.: Polyp-gen: Realistic and diverse polyp image generation for endoscopic dataset expansion. arXiv preprint arXiv:2501.16679, 2025
12. Liu, S., et al.: A comprehensive evaluation of multi-modal large language models for endoscopy analysis. arXiv preprint arXiv:2505.23601 (2025)
13. Klein, H.M.: clinical low field strength magnetic resonance imaging: a practical guide to accessible MRI. Springer (2015)
14. Dou, Q., de Castro, D.C., Kamnitsas, K., Glocker, B.: Domain generalization via model-agnostic learning of semantic features. Advances in neural information processing systems, 32 (2019)
15. Iandola, F.N., Han, S., Moskewicz, M.W., Ashraf, K., Dally, W.J., Keutzer, K.: Squeezenet: Alexnet-level accuracy with 50x fewer parameters and $<$ 0.5 mb model size. arXiv preprint arXiv:1602.07360 (2016)
16. Yu, W., Yang, K., Bai, Y., Xiao, T., Yao, H., Rui, Y.: Visualizing and comparing alexnet and vgg using deconvolutional layers. In: Proceedings of the 33 rd International Conference on Machine Learning, vol. 2 (2016)

17. Kavitha, S., Dhanapriya, B., Naveen Vignesh, G., Baskaran, K.R.: Neural style transfer using vgg19 and alexnet. In: 2021 International Conference on Advancements in Electrical, Electronics, Communication, Computing and Automation (ICAECA), pp. 1–6. IEEE (2021)
18. Islam, K.T., et al.: Improving portable low-field mri image quality through image-to-image translation using paired low-and high-field images. Sci. Rep. **13**(1), 21183 (2023)
19. Dayarathna, S., Islam, K.T., Uribe, S., Yang, G., Hayat, M., Chen, Z.: Deep learning based synthesis of mri, ct and pet: Review and analysis. Med. Image Anal. **92**, 103046 (2024)
20. Dayarathna, S., Islam, K.T., Chen, Z.: Ultra low-field to high-field mri translation using adversarial diffusion. In: 2024 IEEE International Symposium on Biomedical Imaging (ISBI), pp. 1–4. IEEE (2024)
21. Dayarathna, S., et al.: Mccad: multi-contrast mri conditioned, adaptive adversarial diffusion model for high-fidelity mri synthesis. In: 2025 IEEE/CVF Winter Conference on Applications of Computer Vision (WACV), pp. 670–679. IEEE (2025)

Ultra Low-Field MRI Enhancement via Conditional Diffusion Model

Haowen Pang[1], Xueqi Li[2], Tianyi Yan[3], and Chuyang Ye[1(✉)]

[1] School of Integrated Circuits and Electronics, Beijing Institute of Technology, Beijing, China
chuyang.ye@bit.edu.cn
[2] College of Computer Science and Technology, Taiyuan University of Technology, Taiyuan, China
[3] School of Medical Technology, Beijing Institute of Technology, Beijing, China

Abstract. Ultra-low-field magnetic resonance imaging (ULF-MRI) provides a promising low-cost and portable alternative to conventional high-field MRI (HF-MRI), particularly in resource-limited settings. However, its substantially degraded image quality remains a major obstacle to widespread clinical adoption. In this work, we propose a conditional diffusion model (CDM) for enhancing ULF-MRI image quality. During training, multi-modal ULF-MRI inputs (T1W, T2W, and FLAIR) are concatenated with noisy versions of the corresponding HF-MRI images and fed into a U-Net. The U-Net is trained to iteratively predict and remove the noise, thereby progressively synthesizing the HF-MRI. At inference, the multi-modal ULF-MRI inputs are concatenated with Gaussian noise and passed through the trained U-Net, which performs iterative denoising to generate the corresponding HF-MRI. Our framework is trained on paired 64mT–3T MRI data from the ULF-EnC Challenge. Experimental results show that the proposed method substantially improves structural detail and tissue contrast critical for clinical interpretation. These findings highlight the potential of CDM to bridge the quality gap between ULF-MRI and HF-MRI, enabling more accessible and reliable diagnostic imaging in low-resource environments through advanced generative modeling.

Keywords: Low-field MRI · Diffusion model · Image synthesis

1 Introduction

Magnetic resonance imaging (MRI) offering excellent soft-tissue contrast, non-invasiveness, and broad applicability across neurological, oncological, and musculoskeletal domains [6]. However, conventional high-field MRI (HF-MRI), typically operating at 1.5T or 3T, remains prohibitively expensive, immobile, and dependent on specialized infrastructure such as radiofrequency shielding and

H. Pang and X. Li—These authors contributed equally to this work.

Z. Chen et al. (Eds.): ULF-EnC 2025, LNCS 16293, pp. 9–17, 2026.
https://doi.org/10.1007/978-3-032-23344-8_2

cryogenic cooling systems [19]. These requirements hinder widespread deployment in resource-limited environments, including rural clinics, emergency care settings, and developing regions. In contrast, ultra-low-field MRI (ULF-MRI) systems, operating below 0.1T, such as portable 64mT scanners, offer significant advantages in cost, portability, and accessibility. These systems require minimal siting conditions and can be deployed directly at the point of care. Despite their potential, the clinical utility of ULF-MRI is severely constrained by fundamental physical limitations, including low signal-to-noise ratio (SNR), reduced spatial resolution, and poor tissue contrast. These deficiencies hinder accurate interpretation, particularly for tasks demanding fine structural differentiation such as white-gray matter boundary delineation or small lesion detection.

To address this issue, the Ultra-Low-Field MRI Image Enhancement Challenge (ULF-EnC) [1–3,7] provides a public dataset comprising paired 64mT and 3T brain MRI scans across multiple imaging modalities (T1W, T2W, and FLAIR). This initiative aims to encourage the development of advanced enhancement algorithms that map low-field MR images to their high-field counterparts, thereby improving the clinical utility of ULF systems.

Traditional medical image synthesis approaches have primarily relied on convolutional neural networks (CNNs), transformer-based encoder–decoder architectures, or generative adversarial networks (GANs). For instance, Pang et al. employed SwinUNETR [4] for multi-modal brain MRI synthesis, demonstrating the potential of transformer backbones in capturing long-range dependencies while maintaining spatial resolution [10]. Similarly, Zhang et al. [18] developed a CNN–transformer hybrid model for expiratory CT image synthesis, highlighting the complementary strengths of convolutional feature extraction and transformer-based global context modeling. Despite these advances, such methods often face intrinsic limitations. CNN- or transformer-based encoder–decoder frameworks tend to generate blurry reconstructions, especially when synthesizing fine anatomical details that are critical for clinical interpretation. The training of GANs is notoriously unstable, requiring careful tuning and balancing of generator–discriminator objectives to avoid divergence or degraded synthesis quality [12]. These challenges limit the robustness, generalizability, and clinical reliability of conventional medical image synthesis methods, underscoring the need for more stable and anatomically faithful generative paradigms.

Recently, diffusion models (DMs) have emerged as a powerful alternative for image generation, achieving state-of-the-art results in both natural and medical imaging [5,11]. DMs model the data distribution through a forward process that gradually adds noise, and a learned reverse process that iteratively removes it to reconstruct the target image. This gradual generation process offers improved training stability over GANs and allows for more controlled conditioning on auxiliary inputs [8,9,14,16].

In this work, we propose a conditional diffusion model (CDM)-based framework for ULF-MRI enhancement, aiming to synthesize high-field-quality images from multi-modal 64mT scans. The input consists of ULF T1W, T2W, and FLAIR images, concatenated with Gaussian noise at each denoising step and passed to a U-Net-based denoiser. The model learns to reconstruct high-resolution 3T images of the target modality, preserving fine anatomical structures and modality-specific contrasts. We train our model on the official ULF-EnC dataset and design separate networks for synthesizing T1W, T2W, and FLAIR images. Experimental results demonstrate that our approach produces high-quality HF-MRI, substantially improving both perceptual fidelity and structural accuracy. Moreover, the model shows strong generalization across subjects and modalities without relying on adversarial losses or external supervision. The codes of our method are available at https://github.com/PangHaowen-hub/CDM.

2 Methods

2.1 Network Structure

For each target modality (T1W, T2W, or FLAIR), we train a separate generative model that maps multi-contrast ULF-MRI inputs to the corresponding HF-MRI target. As shown in Fig. 1, the core of our framework is a U-Net-based denoiser, which receives as input a four-channel tensor formed by concatenating the noisy image with the three 64mT source modalities (T1W, T2W, and FLAIR). Following [5], the U-Net is based on PixelCNN++ [13] with the Wide ResNet backbone [17]. We adopt the velocity-prediction objective (**v**-prediction), which improves convergence and sample fidelity compared to traditional noise prediction loss.

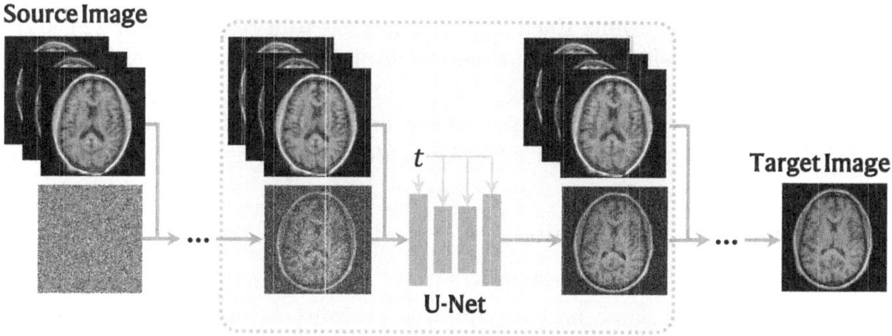

Fig. 1. Overview of the CDM framework. Three-channel 64 mT inputs and Gaussian noise are jointly fed into a U-Net denoiser, which learns to reconstruct high-field quality images across denoising steps.

During the inference stage, we adopt the Denoising Diffusion Implicit Models (DDIM) [15] sampling scheme, which enables deterministic and accelerated generation by skipping a subset of the diffusion steps while maintaining output fidelity. To evaluate the trade-off between synthesis quality and computational efficiency, we experimented with different numbers of denoising steps. Specifically, we compared the 250-step schedule with a reduced 100-step schedule and observed negligible differences in both quantitative metrics and qualitative visual assessments. Consequently, we set 100 inference steps as our default configuration, reducing runtime by approximately 90% compared to the Denoising Diffusion Probabilistic Model (DDPM) [5] with 1000 steps, while preserving image quality. This makes our approach more practical for real-world clinical deployment.

2.2 Data Preprocessing

We utilize the paired MRI dataset provided by the ULF-EnC Challenge [1–3, 7]. For each subject, the dataset includes T1W, T2W, and FLAIR scans acquired at both 64mT and 3T field strengths. To ensure data consistency and usability, a comprehensive preprocessing pipeline was applied. First, all raw scans were converted from the DICOM format to the NIfTI format to facilitate downstream processing. Manual quality control was then conducted to exclude corrupted or incomplete volumes, thereby ensuring that only high-quality data were retained for training and evaluation. Bias field inhomogeneity was corrected using FSL-FAST, which improves intensity uniformity across brain regions. Subsequently, linear registration was performed using FSL-FLIRT to spatially align the 64mT images with their corresponding 3T counterparts. Following registration, all volumes were resampled to an isotropic resolution of 1 mm^3 to eliminate variability in voxel dimensions and to standardize the spatial resolution across the dataset. From each preprocessed volume, we extracted all axial slices without restricting to specific anatomical regions, ensuring that the network is exposed to the full range of brain anatomy. Each slice was resized to a resolution of 224 × 224 pixels. Finally, intensity normalization was applied by rescaling voxel values to the range $[0, 1]$. This preprocessing pipeline ensures both spatial alignment and intensity consistency between the paired low-field and high-field images, thereby creating a reliable foundation for effective model training and robust evaluation.

2.3 Implementation Details

For each of the three target modalities (T1W, T2W, FLAIR), we trained a separate CDM with the same architecture and training setup. The training data includes 49 subjects for training and 1 for validation. Each model was trained using full-slice 2D data, without patch extraction or cropping. During training, we optimize the CDM using the velocity-prediction loss, which minimizes the mean squared error between the predicted velocity vector $\hat{\mathbf{v}}_t$ and the ground-truth velocity vector $\mathbf{v}_t$ at each timestep. The models were trained for 500,000 iterations with a batch size of 16. An exponential moving average with decay rate 0.995 was applied to the model weights to stabilize training. Intermediate samples were generated every 10,000 steps for qualitative monitoring. All training was conducted using PyTorch with automatic mixed precision on a single NVIDIA GPU.

3 Results

3.1 Qualitative Evaluation

We qualitatively assessed the performance of our proposed CDM framework using both our held-out validation subset (sampled from the ULF-EnC training data) and the official ULF-EnC validation set used for competition submission.

Figure 2 shows representative results on our validation subset, illustrating a sample for each target modality (T1W, T2W, and FLAIR). The synthesized HF images (SynHF) were generated from multi-modal ULF-MRIs using our proposed CDM framework. As shown in Fig. 2, the synthesized images exhibit enhanced tissue contrast and improved delineation of anatomical boundaries compared to raw ultra-low-field input. In particular, our method successfully restores subtle structural details that are often obscured in the scans 64 mT. The improved visual quality suggests that the model captures modality-specific intensity distributions and spatial patterns that are characteristic of HF-MRI.

Figure 3 presents representative results on the official ULF-EnC validation set, for which HF ground truth images are not provided. Our model produces images with improved sharpness, contrast, and anatomical coherence compared to the original ULF-MRI. Notably, noise is reduced and deep brain structures appear more distinct, highlighting the robustness and generalizability of our framework.

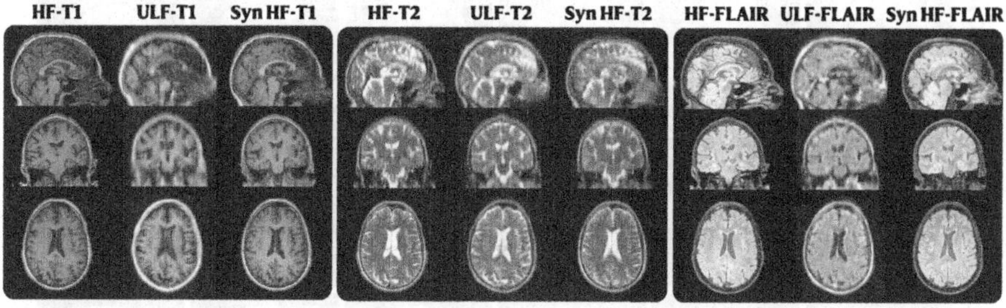

Fig. 2. Results of the proposed CDM on the validation case POCEMR104. For three MRI sequences (T1W, T2W, and FLAIR), HF-MRI, ULF-MRI, and synthesized HF-MRI are shown in three orthogonal views: sagittal, coronal, and axial.

3.2 Quantitative Evaluation

Quantitative evaluation was conducted on both the validation and test sets. For the validation set, the Structural Similarity Index (SSIM), Peak Signal-to-Noise Ratio (PSNR), Mean Absolute Error (MAE), and Normalized Mean Squared Error (NMSE) were computed for the entire image as well as within brain-masked regions, allowing assessment of both global fidelity and anatomical preservation.

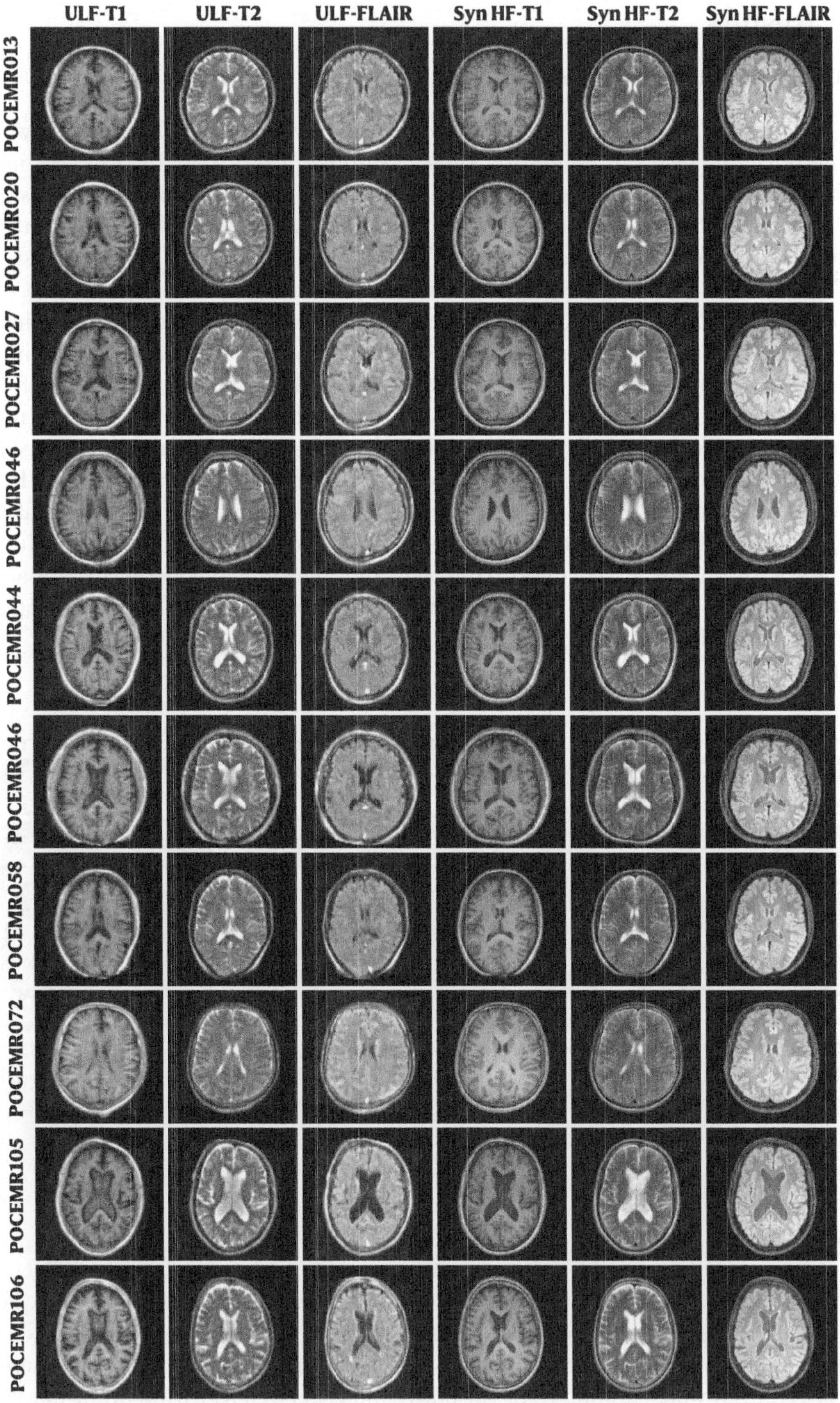

Fig. 3. Qualitative results of the proposed CDM on the ULF-EnC validation set. Each case includes the input ULF-MRI sequences (T1W, T2W, and FLAIR) and the corresponding synthesized HF-MRI.

For the test set, only brain-masked region metrics were reported. The detailed results are summarized in Table 1 and Table 2.

As shown in Table 1, our method achieves an SSIM of 0.810 and PSNR of 22.078 dB for the whole image, while in the brain region the SSIM remains high at 0.645 with a PSNR of 29.722 dB. Low MAE and NMSE values in both evaluation settings indicate accurate intensity recovery and low reconstruction error. These results confirm that our model not only enhances global image appearance but also preserves critical anatomical features necessary for diagnostic interpretation. On the test set (Table 2), our method achieves an SSIM of 0.643 and a PSNR of 29.537 dB within the brain region, accompanied by an MAE of 0.077 and an NMSE of 0.080. The final challenge score reaches 0.727. Taken together, these results confirm that our conditional diffusion model provides reliable ultra-low-field MRI enhancement, balancing global image improvement with fine-grained anatomical preservation critical for clinical deployment.

Table 1. Quantitative evaluation results on the ULF-EnC validation set.

Whole Image				Brain Region			
SSIM	PSNR	MAE	NMSE	SSIM	PSNR	MAE	NMSE
0.810	22.078	0.035	0.083	0.645	29.722	0.076	0.079

Table 2. Quantitative evaluation results on the ULF-EnC test set.

SSIM Masked	PSNR Masked	MAE Masked	NMSE Masked	Final Score
0.643	29.537	0.077	0.080	0.727

4 Discussion

Our study demonstrates the remarkable potential of diffusion-based framework to bridge the quality gap between ULF and HF MRI. The proposed CDM framework effectively transforms low-quality 64mT ULF-MRI data into high-fidelity representations that closely resemble 3T HF-MRI scans. The superior performance of our approach over conventional enhancement baselines stems from its core design principles. First, the underlying methodology is intrinsically built to model and reverse complex noise distributions, a task for which it is exceptionally suited given the degraded signal quality of ULF-MRI. Unlike simpler filtering techniques, our framework does not merely suppress noise; it reconstructs the underlying anatomy by learning to iteratively synthesize information from a noisy state. Second, the iterative denoising process allows the model to synthesize fine structural details and subtle tissue contrasts that are often lost in ULF images. By learning the intricate data distribution from paired 64mT–3T data, our model generates diagnostically relevant features, such as clear gray-white matter differentiation and enhanced visualization of small anatomical structures, which are critical for clinical interpretation.

Despite the promising results, our study has certain limitations that warrant future investigation. First, CDMs are computationally intensive, both during training and inference, due to their iterative nature. While this is less of a concern for offline image processing, real-time applications might require further optimization of the model architecture or faster sampling techniques. Second, our model was trained on the ULF-EnC Challenge dataset, which primarily focuses on healthy brain scans. Further validation is needed to assess its performance and generalizability to various pathologies and different anatomical regions. Third, while our quantitative and qualitative results are encouraging, the ultimate measure of success for medical image enhancement lies in its impact on clinical decision-making. Future work will involve rigorous clinical validation studies with radiologists to assess the diagnostic accuracy and clinical utility of the enhanced ULF images. Future research directions include exploring more efficient CDM variants (e.g., latent diffusion models) to reduce computational overhead, investigating transfer learning approaches to adapt the model to different anatomical sites or pathologies with limited paired data, and incorporating uncertainty quantification to provide clinicians with confidence maps alongside the enhanced images. Additionally, exploring the integration of other complementary imaging modalities or clinical data could further boost the diagnostic value of ULF-MRI.

Disclosure of Interests. The authors have no competing interests to declare that are relevant to the content of this article.

References

1. Dayarathna, S., Islam, K.T., Chen, Z.: Ultra low-field to high-field MRI translation using adversarial diffusion. In: 2024 IEEE International Symposium on Biomedical Imaging (ISBI), pp. 1–4. IEEE (2024)
2. Dayarathna, S., Islam, K.T., Uribe, S., Yang, G., Hayat, M., Chen, Z.: Deep learning based synthesis of MRI, CT and PET: Review and analysis. Med. Image Anal. **92**, 103046 (2024)
3. Dayarathna, S., Islam, K.T., Zhuang, B., Yang, G., Cai, J., Law, M., Chen, Z.: McCaD: multi-contrast MRI conditioned, adaptive adversarial diffusion model for high-fidelity MRI synthesis. In: 2025 IEEE/CVF Winter Conference on Applications of Computer Vision (WACV), pp. 670–679. IEEE (2025)
4. Hatamizadeh, A., Nath, V., Tang, Y., Yang, D., Roth, H.R., Xu, D.: Swin UNETR: Swin transformers for semantic segmentation of brain tumors in MRI images. In: International MICCAI Brainlesion Workshop, pp. 272–284. Springer (2021)
5. Ho, J., Jain, A., Abbeel, P.: Denoising diffusion probabilistic models. Adv. Neural. Inf. Process. Syst. **33**, 6840–6851 (2020)
6. Huijben, E.M., et al.: Generating synthetic computed tomography for radiotherapy: SynthRAD2023 challenge report. Med. Image Anal. **97**, 103276 (2024)

7. Islam, K.T., et al.: Improving portable low-field MRI image quality through image-to-image translation using paired low-and high-field images. Sci. Rep. **13**(1), 21183 (2023)
8. Kazeminia, S., Baur, C., Kuijper, A., van Ginneken, B., Navab, N., Albarqouni, S., Mukhopadhyay, A.: GANs for medical image analysis. Artif. Intell. Med. **109**, 101938 (2020)
9. Kazerouni, A., Aghdam, E.K., Heidari, M., Azad, R., Fayyaz, M., Hacihaliloglu, I., Merhof, D.: Diffusion models in medical imaging: a comprehensive survey. Med. Image Anal. **88**, 102846 (2023)
10. Pang, H., Guo, W., Ye, C.: Multi-modal brain MRI synthesis based on SwinUNETR. arXiv preprint arXiv:2506.02467 (2025)
11. Pang, H., Hong, X., Zhang, P., Ye, C.: Cascaded diffusion model and segment anything model for medical image synthesis via uncertainty-guided prompt generation. In: International Conference on Information Processing in Medical Imaging, pp. 203–217. Springer (2025)
12. Pang, H., et al.: NCCT-CECT image synthesizers and their application to pulmonary vessel segmentation. Comput. Methods Programs Biomed. **231**, 107389 (2023)
13. Salimans, T., Karpathy, A., Chen, X., Kingma, D.P.: PixelCNN++: Improving the PixelCNN with discretized logistic mixture likelihood and other modifications. arXiv preprint arXiv:1701.05517 (2017)
14. Shi, Y., Abulizi, A., Wang, H., Feng, K., Abudukelimu, N., Su, Y., Abudukelimu, H.: Diffusion models for medical image computing: a survey. Tsinghua Sci. Technol. **30**(1), 357–383 (2024)
15. Song, J., Meng, C., Ermon, S.: Denoising diffusion implicit models. In: International Conference on Learning Representations (2021)
16. Wang, T., Lei, Y., Fu, Y., Wynne, J.F., Curran, W.J., Liu, T., Yang, X.: A review on medical imaging synthesis using deep learning and its clinical applications. J. Appl. Clin. Med. Phys. **22**(1), 11–36 (2021)
17. Zagoruyko, S.: Wide residual networks. arXiv preprint arXiv:1605.07146 (2016)
18. Zhang, T.: BreathVisionNet: a pulmonary-function-guided CNN-transformer hybrid model for expiratory CT image synthesis. Comput. Methods Programs Biomed. **259**, 108516 (2025)
19. Zhao, Y., Ding, Y., Lau, V., Man, C., Su, S., Xiao, L., Leong, A.T., Wu, E.X.: Whole-body magnetic resonance imaging at 0.05 Tesla. Science **384**(6696), eadm7168 (2024)

Utilizing an Ensemble of 3D U-Nets to Predict High-Field MRI T1-, T2-, and FLAIR-Images from Ultra-Low-Field MRI Images

Jan Nikolas Morshuis[1(✉)], Matthias Hein[1], and Christian F. Baumgartner[1,2]

[1] University of Tübingen, Tübingen, Germany
nikolas.morshuis@uni-tuebingen.de
[2] University of Lucerne, Lucerne, Switzerland

Abstract. Low-field Magnetic Resonance Imaging (MRI) has the potential to provide the benefits of MRI diagnosis like high soft tissue contrast to regions of the world that do not have the required infrastructure in order to run the more expensive high-field MRI machines. However, due to the low magnetic field strengths that the low-field scanners have, the image quality is usually worse compared to the high-field scanners and less details can be recognized. In recent years, deep learning methods have shown to be able to reconstruct images of accelerated MRI. In accelerated MRI - similar to low-field MRI - it is also the goal to reconstruct the original MRI image from the accelerated MRI image, that usually is degraded compared to the original. While many deep learning based approaches exist in the area of MRI reconstruction, one common well-functioning baseline is to utilize 3-dimensional U-Nets for the reconstruction task. In this paper we will explore how we can utilize ensembles of U-Nets in order to predict high-field MRI images from low-field MRI images. We show that including more U-Nets in the ensemble can increase the SSIM and PSNR scores of the predictions compared with the high-field MRI ground-truth.

Keywords: Low-field MRI · MRI-Reconstruction · Image Regression

1 Introduction

Magnetic Resonance Imaging (MRI) is an essential tool in modern diagnostic radiology and clinical neuroscience, offering high-resolution, non-invasive insights into soft tissue structures such as the brain, spinal cord, and internal organs. High-field MRI scanners, typically operating at field strengths of 1.5 T or higher, are the standard in many hospitals and imaging centers due to their superior signal-to-noise ratio (SNR), contrast, and spatial resolution. These qualities make high-field MRI indispensable for accurate diagnosis, surgical planning, and longitudinal patient monitoring. However, these systems come with significant financial and infrastructural demands. High-field MRI scanners

Z. Chen et al. (Eds.): ULF-EnC 2025, LNCS 16293, pp. 18–27, 2026.
https://doi.org/10.1007/978-3-032-23344-8_3

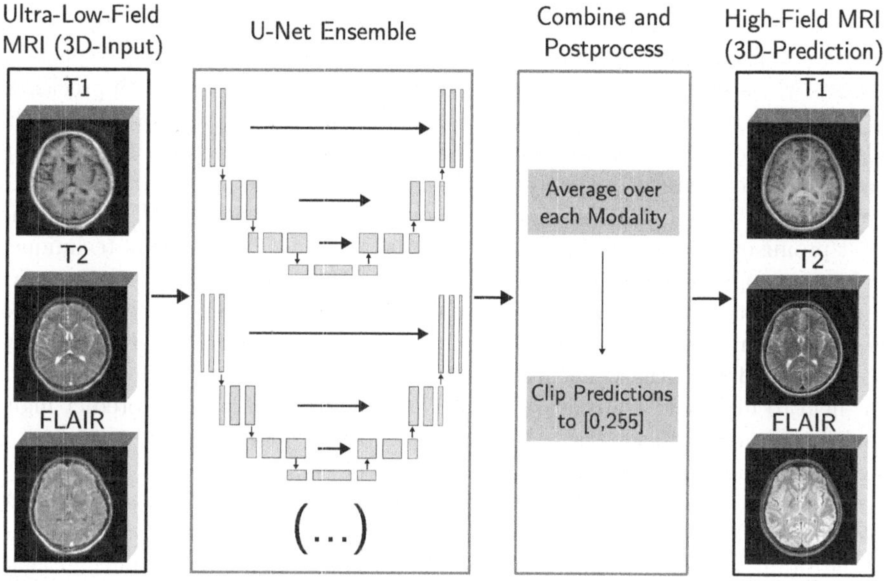

Fig. 1. An overview of our 3D U-Net ensemble for ultra-low-field to high-field MRI translation. Our framework leverages an ensemble of 3D U-Nets to process the concatenated T1, T2, and FLAIR modalities. The individual model outputs are fused via averaging and clipped to a [0, 255] intensity range. The resulting high-field images demonstrate a significant enhancement in anatomical detail and contrast compared to the ultra-low-field inputs.

are extremely expensive to acquire and install, often costing several million dollars. Furthermore, their operation requires advanced cooling systems, electromagnetic shielding, continuous and stable electricity supply, and highly trained technical staff for maintenance and operation. As a result, access to high-field MRI remains limited, particularly in low- and middle-income countries, rural areas, and under-resourced healthcare environments.

In contrast, low-field MRI scanners, which typically operate at field strengths as low as 64mT, present a more affordable and logistically feasible alternative. These systems are more compact, consume significantly less power, and can be deployed in environments where high-field MRI is not viable. Recent advances in hardware design and image acquisition techniques have further improved the usability of low-field MRI. Notably, some low-field scanners are portable and can be moved between hospital wards or even used in remote locations. Their reduced operational costs make them particularly attractive for expanding the global accessibility of diagnostic imaging [1]. These advantages are especially relevant for ultra-low-field scanners in the millitesla range [2].

Despite these advantages, low-field MRI suffers from inherent limitations due to its reduced magnetic field strength [3]. These include lower SNR, reduced tissue contrast, longer acquisition times, and diminished resolution, all of which can lead to reduced diagnostic utility compared to high-field MRI. In clinical practice, this can result in ambiguous findings or missed pathologies, especially in cases where subtle structural anomalies must be detected with high precision.

To bridge this quality gap, there is a growing interest in computational methods that can enhance low-field MRI data. This is part of a broader field of deep learning-based medical image synthesis, which has shown significant success in translating between different imaging modalities and qualities [4]. One of the most promising approaches is to leverage image-to-image translation techniques to learn mappings from low-field to high-field MRI images, effectively improving the quality of portable low-field scans [5]. Recent work has demonstrated the particular power of advanced generative models, such as adversarial diffusion models, for synthesizing high-fidelity medical images and performing ultra low-field to high-field translation [6,7]. These models can learn to predict high-field image quality from low-field scans, effectively simulating the output of a high-field scanner without requiring the physical presence of one.

The "Ultra-Low-Field MRI Image Enhancement Challenge 2025" (ULF-EnC) addresses precisely this task: Given low-field T1-, T2- and FLAIR-MRI images of the brain, the goal is to reconstruct or predict the corresponding high-field scan as accurately as possible. This is a highly non-trivial problem that requires models to learn complex anatomical correspondences, compensate for noise and resolution loss, and generate outputs that are both perceptually convincing and quantitatively accurate.

In this paper, we propose a deep learning-based approach to this task, focusing on producing high-quality reconstructions that faithfully reproduce the structural details of the target high-field MRI images. We describe our preprocessing pipeline, model architecture, and training strategy in detail, and demonstrate the effectiveness of our approach on the challenge dataset. Our method achieves competitive performance while maintaining robustness across patients and imaging conditions. By contributing a strong and generalizable baseline to this problem, we hope to facilitate broader access to high-quality neuroimaging and stimulate further research in the area of low-to-high-field MRI translation.

2 Method

Our approach to predicting high-field MRI images from low-field inputs is based on an ensemble of convolutional neural networks, each adopting a U-Net architecture [8]. Our method is designed to leverage the complementary information available in multiple MRI sequences (T1-weighted, T2-weighted, and FLAIR) and to robustly generate anatomically faithful high-field reconstructions of the same modalities. An overview of our method is shown in Fig. 1.

2.1 Overview

Given the input low-field scans for a subject – specifically, the T1-weighted, T2-weighted, and FLAIR sequences – our model predicts the corresponding high-field scans for the same three modalities. Rather than building separate models for each modality, we opt for a multi-input, multi-output framework that processes all three low-field inputs jointly and predicts the three high-field modalities simultaneously. This joint modeling approach allows the network to exploit cross-modal dependencies and anatomical consistency across sequences.

To further improve the robustness and generalization of the method, we employ an ensemble of multiple U-Net models, each trained independently with different random seeds and data augmentations. At inference time, the predictions from each model in the ensemble are averaged to produce the final output.

2.2 Input and Output Representation

Each subject in the challenge dataset is represented by three co-registered volumetric MRI scans acquired with a ultra-low-field MRI scanner: one T1-weighted scan, one T2-weighted scan, and one FLAIR scan. The data was acquired using a Siemens Biograph mMR (3T) scanner and a Hyperfine Swoop (64mT) scanner. The volumes stacked along the channel dimension to form a single three-channel input volume.

The model predicts a corresponding three-channel output volume, where each channel represents the predicted high-field version of the T1, T2, and FLAIR sequences, respectively. This design ensures a consistent mapping between input and output modalities and facilitates joint optimization across channels.

2.3 Network Architecture

We orientate our approach on the regression network introduced by the winning method of the K2S-challenge [9] and that was also utilized in [10], where a simple 3D U-Net reconstruction achieved the best reconstruction scores for accelerated MRI. For each model in the ensemble, we adopt a standard 3D U-Net framework architecture as provided by the nnU-Net framework [11], which has been widely used in medical image analysis due to its ability to model both local and global context. All convolutional operations are three-dimensional to fully exploit the volumetric nature of the given MRI data.

2.4 Ensemble Strategy

To increase performance and reduce overfitting, we train each ensemble with $N = 5$ U-Net models. Each model is initialized with a different random seed and trained with stochastic data augmentations. Each model only trains on 40 out of the 50 available volumes for training and utilizes the remaining 10 volumes for validation during training. This ensemble strategy introduces diversity among the models and improves the overall robustness of the predictions.

At inference time, the ensemble members each generate their own prediction for a given input volume. The final output is computed as the voxel-wise average of these predictions. This simple averaging technique reduces variance and helps to correct individual prediction errors, leading to smoother and more accurate reconstructions.

2.5 Loss Function

The models are trained to minimize a voxel-wise Mean-Squared-Error (MSE) and the Structural Similarity Index Measure (SSIM) between the predicted high-field outputs and the corresponding ground truth high-field images. During training, the loss function combines both losses as shown in Eq. 1:

$$\mathcal{L}_{\text{total}} = \lambda_{\text{MSE}} \cdot \mathcal{L}_{\text{MSE}} + \lambda_{\text{SSIM}} \cdot (1 - \text{SSIM}) \quad (1)$$

where λ_{MSE} and λ_{SSIM} are hyperparameters controlling the relative contribution of each term. MSE encourages voxel-wise fidelity, while SSIM promotes the preservation of structural and perceptual image quality. We empirically choose $\lambda_{\text{MSE}} \approx \lambda_{\text{SSIM}} \approx 0.5$.

2.6 Training Details

We utilize the nnU-Net framework [11] for the training of our networks. The nnU-Net is originally created for medical image segmentation, yet it's specialization on 3D-medical images is also useful for the reconstruction task that needs to be solved for the challenge at hand.

For our final method, we developed a dual-normalization ensemble strategy (Dual-Ensemble U-Nets) to maximize model robustness and performance. This approach combines models trained on two distinct data normalization schemes. The first is min-max normalization, which scales the challenge data's voxel values from the range [0, 255] to [0, 1]. The second is z-score normalization (to zero mean and unit variance), which is advantageous for small brain regions with extreme intensity values, as errors in these areas are penalized more heavily by the MSE loss function. Our final model consists of two 5-model U-Net ensembles: one trained on min-max normalized data and the other on z-score normalized data. We combine both ensembles using a weighted average, giving the ensemble trained on the min-max scaled data a weight of 0.9 and the ensemble trained on z-score normalized data a weight of 0.1. All predictions are rescaled using the respective inverse method to [0,255] and all values < 0 or > 255 are clipped.

All models are trained on 3D patches randomly extracted from the full MRI volumes to reduce memory consumption and increase data diversity. The original size of the MRI images is $160 \times 224 \times 220$. Patches of size $112 \times 160 \times 128$ voxels are sampled uniformly across the brain volume. We try to keep the patch size as large as possible with the given compute, because larger patch-sizes allow to correct for potential registration errors.

We use a batch size of 2. Each model is trained for 1000 epochs of 250 iterations. The training is performed using full-precision on a single NVIDIA A100 GPU.

3 Results

In this section, we report and analyze the performance of our proposed method on the validation set provided by the challenge organizers. We evaluate our models using the standard quantitative metrics made available on the Synapse platform, and we complement these findings with qualitative visualizations to assess the anatomical plausibility and perceptual quality of the predicted high-field MRI images.

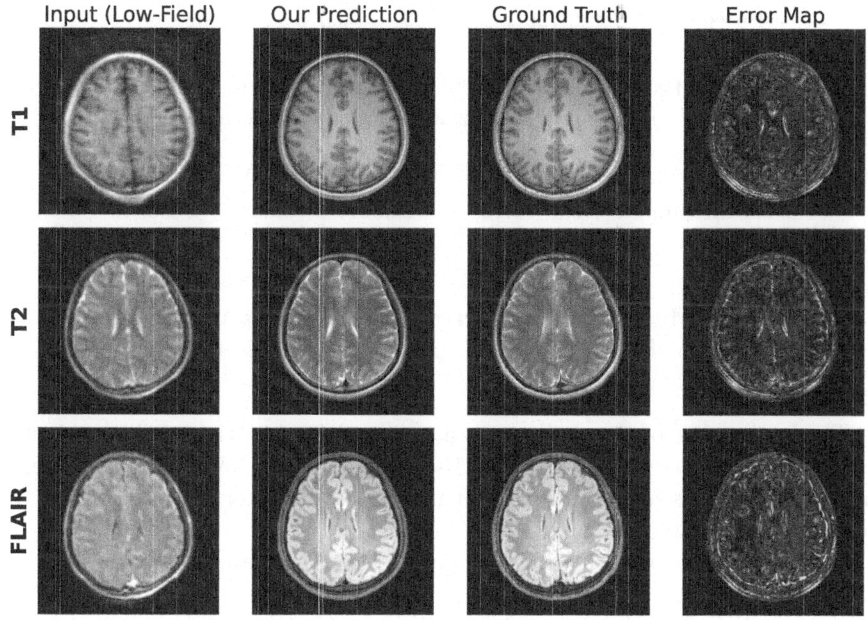

Fig. 2. Qualitative results: Our method is able to generate MRI images that appear similar to the high-field MRI Ground Truth. The error is generally rather low, but is higher close to the skull or the lateral ventricles. This image has been generated on the validation set of a single fold.

3.1 Evaluation Metrics

The challenge evaluates predictions using several commonly used image similarity metrics, including:

- **Structural Similarity Index (SSIM)**: Assesses perceptual similarity between predicted and target images, accounting for luminance, contrast, and structure.
- **Peak Signal-to-Noise Ratio (PSNR)**: Quantifies the ratio between the maximum possible signal and the noise in the prediction.
- **Mean Absolute Error (MAE)**: Measures the average absolute voxel-wise difference between the predicted and ground truth high-field images.

Table 1. Validation results from the Synapse platform for our submitted methods. Note that all models during the validation phase have been trained with z-score normalization. The ensemble clearly performs better compared to the single U-Net.

Model	SSIM-M ↑	PSNR-M ↑	MAE-M ↓	NMSE-M ↓
Single U-Net	0.690	27.169	0.077	0.071
Ensemble U-Nets	0.720	28.599	0.070	0.062

Table 2. Test results from the Synapse platform of our final method, combining both ensembles trained with min-max normalization and z-score normalization

Model	SSIM-M ↑	PSNR-M ↑	MAE-M ↓	NMSE-M ↓
Dual-Ensemble U-Nets	0.728	29.821	0.068	0.064

- **Normalized Mean Squared Error (NMSE)**: Measures the squared difference between the predicted and ground truth images, normalized by the energy of the ground truth.

The final metric of the challenge is a weighted combination of all the masked versions of these losses (SSIM-M, PSNR-M, MAE-M and NMSE-M), that means that only the parts of the image belonging to the brain or the skull region are considered for the score evaluations.

The scores are combined using a weighted average, namely

$$0.7 \cdot SSIM + 0.1 \cdot PSNR + 0.1 \cdot MAE + 0.1 \cdot NMSE \tag{2}$$

3.2 Quantitative Results

Table 1 and Table 2 show the performance of our tested models on the validation set offered by the challenge. The results are extracted directly from the Synapse leaderboards and reflect the average performance across all subjects and modalities. Note that during the model validation timeslot only models trained with the z-score normalization have been submitted.

Note that large improvements exist between the single U-Net and the Ensemble U-Nets according to Table 1. As the task of predicting the high-field MRI image from the low-field MRI images might have an inherent uncertainty, it might not be possible to predict all structure of the high-field MRI brain images precisely. The combination of several possible reconstructions therefore smoothes the image in areas that are uncertain and therefore potentially lead to better PSNR scores. These results clearly support our hypothesis that ensembling contributes to improved generalization and more stable predictions. The ensemble not only reduces voxel-level error but also enhances perceptual similarity, as evidenced by higher SSIM scores.

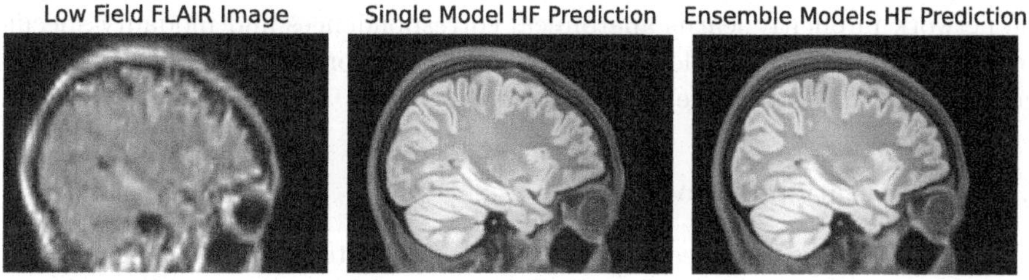

Fig. 3. Qualitative results for one subject for a FLAIR MRI image. Note that in the low-field MRI image, most of the structure is difficult to recognize. As our models use all three input channels available (T1, T2, and FLAIR), the brain structure can likely be inferred from the other input modalities. Also note that the difference between the Single Model prediction and Ensemble Models prediction is subtle.

3.3 Qualitative Results

In addition to the numerical results, we visualize a subset of predicted volumes to assess the perceptual quality and anatomical fidelity of the reconstructions. Figure 2 shows an axial slice from a representative subject in the validation set, comparing the low-field input with a prediction of a single model and the ground truth. We also show a sagittal slice for the FLAIR modality in Fig. 3.

As shown, our model captures fine anatomical structures. The contrast and sharpness of the predicted images are more detailed and closer resemble the ones of an MRI image that is acquired using a high-field MRI scanner.

4 Discussion

In this section, we reflect on the performance and behavior of our proposed method, with a particular emphasis on the impact of using an ensemble of U-Net models as opposed to relying on individual predictions. Our results clearly indicate that ensembling contributes substantially to the overall prediction quality in terms of standard quantitative metrics.

The ensemble acts as a form of implicit regularization. By combining multiple hypotheses, it produces predictions that are more conservative and stable, particularly in regions where the low-field input offers limited or ambiguous information. This robustness is critical for clinical applications, where trustworthiness and consistency are paramount.

Moreover, although not explored quantitatively in this work, the ensemble could be further leveraged for uncertainty estimation. For example, the voxel-wise variance across the ensemble members' outputs could be interpreted as a proxy for epistemic uncertainty, which could be valuable in downstream clinical decision-making pipelines, as suggested by Küstner et al. [12].

An additional benefit of joint multi-modal prediction within the ensemble framework is the improved consistency across output modalities. Because the three high-field modalities (T1, T2, and FLAIR) are predicted simultaneously, the network implicitly learns anatomical correspondences and modality-specific contrasts. This multi-task learning setup acts as a form of inductive bias, promoting structural coherence across outputs.

4.1 Computational Trade-Offs

It is important to acknowledge the trade-offs associated with the use of ensembling. Training multiple models and storing their weights incurs additional computational and memory costs. Inference time is also proportionally longer, since predictions must be generated for each model in the ensemble before averaging.

4.2 Limitations and Future Directions

While the ensemble approach delivers strong results, it is not without limitations. Future work could investigate more sophisticated ensemble construction techniques, such as using models with different architectures, training regimes, or data subsets to further increase the diversity and coverage of the ensemble.

Acknowledgments. Funded by the Deutsche Forschungsgemeinschaft (DFG, German Research Foundation) under Germany's Excellence Strategy - EXC number 2064/1 - Project number 390727645. The authors thank the International Max Planck Research School for Intelligent Systems (IMPRS-IS) for supporting JNM.

Disclosure of Interests. The authors have no competing interests to declare that are relevant to the content of this article.

References

1. Campbell Arnold, T., Freeman, C.W., Litt, B., Stein, J.M.: Low-field mri: clinical promise and challenges. J. Magnetic Resonance Imaging **57**(1), 25–44 (2023)
2. Sarracanie, M., Salameh, N.: Low-field mri: how low can we go? a fresh view on an old debate. Front. Phys. **8**, 172 (2020)
3. Marques, J.P., Simonis, F.F.J., Webb, A.G.: Low-field mri: an mr physics perspective. J. Magnet. Resonance Imaging **49**(6), 1528–1542 (2019)
4. Dayarathna, S., Islam, K.T., Uribe, S., Yang, G., Hayat, M., Chen, Z.: Deep learning based synthesis of mri, ct and pet: Review and analysis. Med. Image Anal. **92**, 103046 (2024)
5. Islam, K.T., et al.: Improving portable low-field mri image quality through image-to-image translation using paired low-and high-field images. Sci. Rep. **13**(1), 21183 (2023)
6. Dayarathna, S., Tohidul Islam, K., Chen, Z.: Ultra low-field to high-field mri translation using adversarial diffusion. In: 2024 IEEE International Symposium on Biomedical Imaging (ISBI), pp. 1–4. IEEE (2024)
7. Dayarathna, S., et al.: Mccad: multi-contrast mri conditioned, adaptive adversarial diffusion model for high-fidelity mri synthesis. In: 2025 IEEE/CVF Winter Conference on Applications of Computer Vision (WACV), pp. 670–679. IEEE (2025)

8. Ronneberger, O., Fischer, P., Brox, T.: U-net: convolutional networks for biomedical image segmentation. In: International Conference on Medical Image Computing and Computer-Assisted Intervention, pp. 234–241. Springer (2015)
9. Tolpadi, A.A., et al.: K2s challenge: from undersampled k-space to automatic segmentation. Bioengineering **10**(2), 267 (2023)
10. Morshuis, J.N., Hein, M., Baumgartner, C.F.: Understanding benefits and pitfalls of current methods for the segmentation of undersampled mri data (2025)
11. Isensee, F., Jaeger, P.F., Kohl, S.A.A., Petersen, J., Maier-Hein, K.H.: nnu-net: a self-configuring method for deep learning-based biomedical image segmentation. Nature Methods **18**(2), 203–211 (2021)
12. Küstner, T., Hammernik, K., Rueckert, D., Hepp, T., Gatidis, S.: Predictive uncertainty in deep learning–based mr image reconstruction using deep ensembles: Evaluation on the fastmri data set. Magn. Reson. Med. **92**(1), 289–302 (2024)

Augment to Augment: Diverse Augmentations Enable Competitive Ultra-Low-Field MRI Enhancement

Felix F. Zimmermann(✉)

Physikalisch-Technische Bundesanstalt (PTB), Braunschweig and Berlin, Germany
felix.zimmermann@ptb.de

Abstract. Ultra-low-field (ULF) MRI promises broader accessibility but suffers from low signal-to-noise ratio (SNR), reduced spatial resolution, and contrasts that deviate from high-field standards. Image-to-image translation can map ULF images to a high-field appearance, yet efficacy is limited by scarce paired training data. Working within the ULF-EnC challenge constraints (50 paired 3D volumes; no external data), we study how task-adapted data augmentations impact a standard deep model for ULF image enhancement. We show that strong, diverse augmentations, including auxiliary tasks on high-field data, substantially improve fidelity. Our submission ranked third by brain-masked SSIM on the public validation leaderboard and fourth by the official score on the final test leaderboard.

Code is available at https://github.com/fzimmermann89/low-field-enhancement.

Keywords: ULF-EnC · ultra-low-field MRI · image enhancement

1 Introduction

Magnetic Resonance Imaging (MRI) is a cornerstone of medical diagnostics, with clinical practice dominated by systems operating at high field strengths of 1.5 T and 3 T. These systems rely on sophisticated superconducting magnets, rendering them large, costly, and complex to site. In contrast, ultra-low-field (ULF) MRI, which operates at field strengths below 100 mT, has emerged as a promising alternative [1]. The potential advantages of ULF MRI include reduced cost, enhanced safety, and fewer image artifacts from metallic implants or foreign bodies.

However, these benefits are counterbalanced by a fundamental physical limitation: the signal amplitude is proportional to the magnetic field strength, leading to an inherently low signal-to-noise ratio (SNR) in ULF images [1,2]. This low SNR manifests as reduced achievable spatial resolution and increased image noise. Furthermore, the field-dependent nature of tissue relaxation mechanisms results in image contrasts that differ significantly from established clinical standards, posing an interpretative challenge for radiologists.

Z. Chen et al. (Eds.): ULF-EnC 2025, LNCS 16293, pp. 28–36, 2026.
https://doi.org/10.1007/978-3-032-23344-8_4

Efforts to overcome the limitations of ULF MRI are advancing on two primary fronts. The first involves innovations in hardware and data acquisition, including the design of novel acquisition schemes and improvements in signal amplification and noise rejection. The second front, which is the focus of this work, concentrates on image reconstruction and post-processing [1,3–6]. This approach leverages computational imaging techniques and deep learning to mitigate the effects of low SNR and bridge the contrast gap to high-field standards.

Deep learning for ULF enhancement proceeds via two broad strategies. Physics-informed learned reconstruction integrates networks into the image formation from k-space with data consistency [7–9], an approach that is state-of-the-art for accelerated high-field MRI [10–13]. This approach requires access to raw data, which is commonly unavailable on commercial systems. Alternatively, post-processing performs image-to-image translation on conventional reconstructions [14,15], which we adopt.

We operate within the ULF-EnC challenge [16] setting: 50 paired 3D brain volumes acquired at 64 mT (Hyperfine Swoop) and 3 T (Siemens Biograph mMR) for T1-weighted, T2-weighted, and FLAIR, co-registered for paired training and evaluation. Our submission ranked third by brain-masked SSIM on the public validation leaderboard and ultimately placed fourth on the final test leaderboard.

2 Methods

Our method for low- to high-field MRI translation is based on three key components: a multi-task network architecture, a hybrid training objective, and a strong data augmentation strategy, as illustrated in Fig. 1.

The full source code and trained weights as submitted to the challenge are available at https://github.com/fzimmermann89/low-field-enhancement.

The model was trained for 500 epochs using AdamW [17] with a cosine learning rate schedule, taking approximately 24 h on four NVIDIA H100 GPUs.

2.1 Network Architecture

A single multi-contrast 3D U-Net ingests T1-weighted, T2-weighted, and FLAIR volumes as channels. The target contrast and task is provided via an embedding and used for FiLM conditioning (c in Fig. 1). The architecture follows a standard 3D U-Net [18] as implemented in MRpro [19] with four resolution levels with 64→128→192→256 channels. Each encoder/decoder level comprises two ResNet blocks with two convolutions each, SiLU activation, Group Normalization [20], and FiLM [21] in the ResNet blocks and the skip connections. Multi-head self-attention is inserted at the two lowest resolutions for long-range context. We use strided convolutions for downsampling and trilinear interpolation for upsampling. A coordinate grid is concatenated to the inputs of the first convolution to provide explicit spatial context [22]. The model has approximately 90 M parameters.

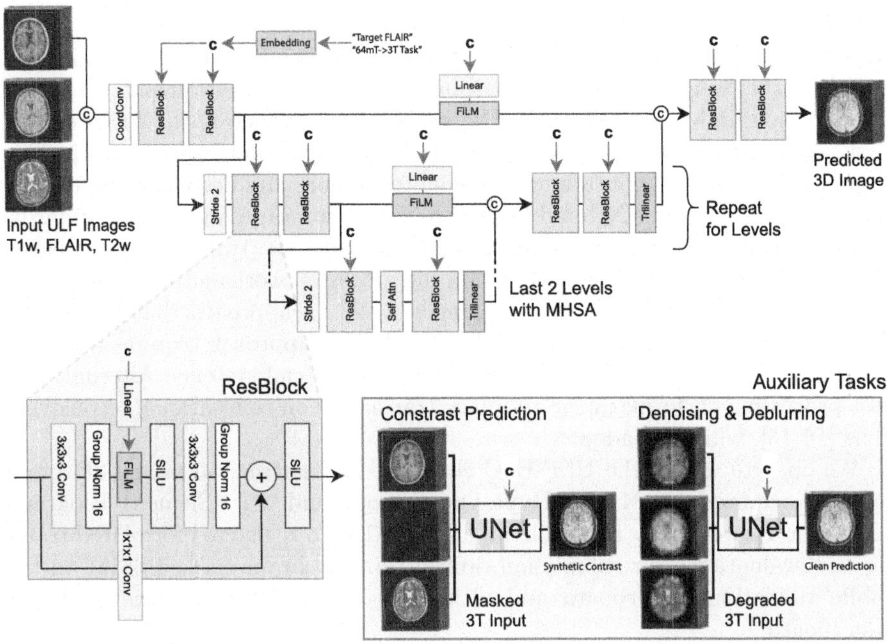

Fig. 1. Schematic overview of the proposed method. A single 3D U-Net is trained on three tasks. The primary task (top) is to translate the three low-field contrasts into a target high-field contrast, specified via FiLM conditioning. Two auxiliary tasks augment the training by leveraging only the high-field data: (1) contrast synthesis and (2) image restoration (denoising/deblurring).

2.2 Training Objective

The training objective combines a supervised reconstruction loss with an adversarial loss [5,23,24]. The reconstruction loss is a weighted sum of a pixel-wise L1 loss and a 3D structural similarity (SSIM) loss (rectangular $11 \times 11 \times 11$ window), with weights 0.2 and 0.8, respectively. To mitigate the blurring often associated with pixel-wise losses, we incorporate an adversarial loss with weight 0.2. This is implemented using a conditional PatchGAN [25] discriminator (4 Levels of stride 2, $4 \times 4 \times 4$ kernel convolutions, 32→64→128→256 features, LeakyReLU activations, GroupNorm) within a hinge loss formulation [26]. Conditioning on the target sequence is achieved by embedding the target into an 8-dimensional vector, broadcasting, and concatenating in the channel dimension. The discriminator is regularized with an R1 gradient penalty [27] every second step.

2.3 Auxiliary Tasks

In addition to the primary task of low-to-high field translation, we train the network on two auxiliary tasks that leverage only the provided high-field data:

high-field contrast synthesis and high-field image restoration [28]. To accommodate these, the network input is designed to accept both low-field and high-field volumes in separate channels. For the main translation task, the high-field input channels are zeroed; for the auxiliary tasks, the low-field input channels are zeroed. For contrast synthesis, the network must learn to recreate a target contrast from the other two available high-field contrasts; this is achieved by zeroing out the target contrast channel in the input [29]. For image restoration, the network is trained as a denoiser and deblurrer on randomly degraded high-field images.

2.4 Data Augmentations

Given the limited training data and the challenge's prohibition on external data, a robust augmentation strategy is essential [30,31]. Our approach integrates auxiliary training tasks with a diverse set of standard data augmentations.

Geometric Augmentations. We apply a set of random geometric augmentations consistently across the input contrasts and the target volume. These include standard affine transformations (rotations, shifts, shearing), left-right flips, and moderate 3D non-rigid transformations.

Intensity Augmentations. To improve robustness against variations in input contrast, a random monotonic intensity mapping is applied to each input volume in 20% of training samples. For each contrast, four support values are randomly sampled from the range $[0, 1]$. These values are sorted and then assigned as the output intensities corresponding to the fixed input intensities of 0.2, 0.4, 0.6, and 0.8. A complete, continuous mapping is generated via linear interpolation between these support points, creating artificial variations in the input image contrast.

Input Degradations. To further increase robustness, the input volumes are additionally degraded in 20% of training samples [32]. This involves adding Gaussian noise with a small, randomly selected standard deviation to each input contrast independently. Furthermore, we apply anisotropic blurring with a randomized kernel strength for each spatial dimension.

3 Results

Two training subjects were held out for internal validation. Exemplary central axial slices from one validation subject are shown in Fig. 2 on the facing page. In all three contrasts, the predicted enhanced images are visually close to the target 3 T images. The SSIM values for this volume are 0.79/0.82/0.83 for the FLAIR/T1-/T2-weighted contrasts. Official challenge results on the official validation and test sets are reported in Table 1 on page 6. The score used by the challenge, a combination of different metrics

$$\text{Score} = 0.7\,\text{SSIM} + 0.1\,\frac{\text{PSNR}}{32.0} + 0.1\,(1 - \text{MAE}) + 0.1\,(1 - \text{NMSE})$$

was 0.779, placing our submission 4th out of 26 submissions from 18 teams.

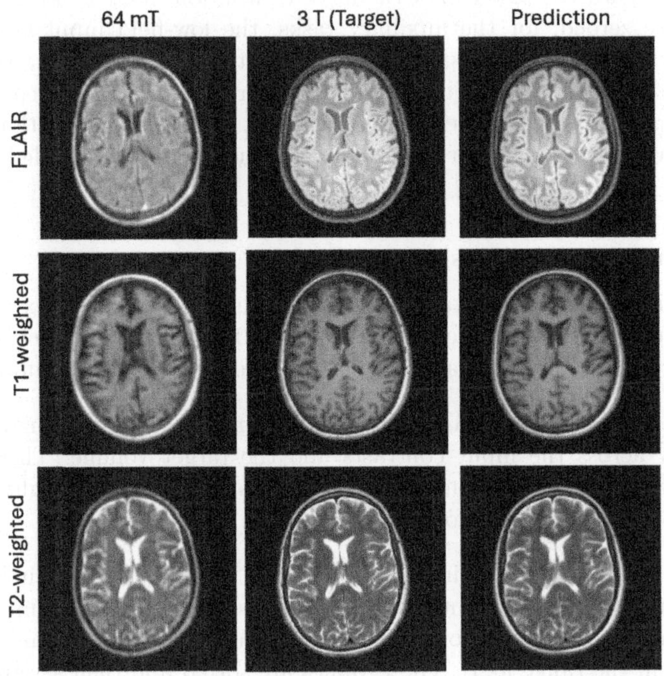

Fig. 2. Exemplary result of the central axial slice of a held-out training sample. Given FLAIR (top), T1-weighted (center), and T2-weighted (bottom) input volumes (left), our trained network predicts 3 T-like image volumes (right), closely matching the ground truth targets (center).

3.1 Ablation Study

We investigate the influence of different components of our method. The ablations were chosen to isolate the impact of our augmentation strategy (a, b, c), the choice of loss function (d), and the 3D network architecture (e):

a) *No Augmentations*: All augmentations and auxiliary tasks removed.
b) *Only Affine Augmentations*: Only standard random affine transformation and flips.
c) *No Auxiliary Tasks*: All data augmentations, but no auxiliary tasks.
d) *No Adversarial Loss*: Training with only the L1+SSIM reconstruction loss.
e) *2D U-Net*: A 2D U-Net trained on axial slices instead of the proposed 3D U-Net.

Table 1. Results of our proposed method on the validation and test set as reported by the official leaderboards.

Metric	Validation		Test
	Value	Value Masked	Value Masked
SSIM	0.823	0.715	0.714
PSNR	23.02	30.01	29.84
MAE	0.032	0.069	0.070
NMSE	0.067	0.153	0.066

The results on the internal validation samples, shown in Table 2 on the following page, highlight the substantial performance benefit of our diverse augmentation strategy. The 3D approach and the adversarial loss also prove to be critical components.

Table 2. Ablation study on held-out training subjects (internal validation).

Ablation	SSIM
a) No Augmentation	0.68
b) Only Affine Augmentations	0.76
c) No Auxiliary Tasks	0.79
d) No Adversarial Loss	0.80
e) 2D U-Net	0.72
Proposed	**0.82**

4 Discussion

We presented a deep learning method for ULF MRI enhancement that combines a contrast-conditioned 3D U-Net, a hybrid reconstruction–adversarial objective, and a diverse augmentation strategy. The ablation study confirmed the importance of our multi-faceted augmentation approach for achieving high performance with limited data. Challenge results independently demonstrate the competitiveness of this strategy.

Our design choices were guided by observations of the training data. We noted that for a significant number of training samples, the co-registration between the 3 T and 64 mT volumes appeared suboptimal. This motivated our decision to use a 3D, multi-contrast approach, where all three source contrasts are used to predict a single target volume, allowing the network to leverage robust anatomical information from one contrast to aid the reconstruction of another.

A general limitation of learned image enhancement is the difficulty in assessing model uncertainty. The enhanced images may appear sharp and convincing even when generated from poor-quality inputs, potentially masking uncertainty or hallucinating anatomical details. This ill-posed problem is a critical barrier to clinical translation. To address this, we briefly investigated a diffusion-based model capable of sampling from the posterior distribution of possible high-field images. While this approach effectively visualizes model uncertainty by generating multiple, distinct high-field reconstructions for a single input, its performance on standard metrics was not competitive within the challenge constraints. We hypothesize this can be addressed by using additional, unpaired public training data and plan to pursue this avenue further.

Acknowledgments. This work was supported in part by the Metrology for Artificial Intelligence for Medicine (M4AIM) Project that is funded by the German Federal Ministry for Economic Affairs and Climate Action (BMWi) in the framework of the QI-Digital initiative and in part by the Deutsche Forschungsgemeinschaft (DFG, German Research Foundation) under Grant 372486779 (SFB1340).

Disclosure of Interests. The authors have no competing interests to declare that are relevant to the content of this article.

References

1. Arnold, T.C., Freeman, C.W., Litt, B., Stein, J.M.: Low-field MRI: clinical promise and challenges. J. Magn. Reson. Imaging **57**(1), 25–44 (2023)
2. Marques, J.P., Simonis, F.F.J., Webb, A.G.: Low-field MRI: an MR physics perspective. J. Magn. Reson. Imaging **49**(6), 1528–1542 (2019)
3. Schote, D., et al.: Physics-informed deep learning for image distortion correction from B 0-inhomogeneities in Low-Field MRI. In: Proc. Intl. Soc. Mag. Reson. Med., vol. 31 (2022)
4. Ayde, R., Vornehm, M., Zhao, Y., Knoll, F., Wu, E.X., Sarracanie, M.: MRI at low field: a review of software solutions for improving SNR. NMR Biomed. **38**(1), e5268 (2025)
5. Dayarathna, S., Islam, K.T., Chen, Z.: Ultra low-field to high-field MRI translation using adversarial diffusion. In: 2024 IEEE International Symposium on Biomedical Imaging (ISBI), pp. 1–4. IEEE (2024)
6. Dayarathna, S., Islam, K.T., Uribe, S., Yang, G., Hayat, M., Chen, Z.: Deep learning based synthesis of MRI, CT and PET: review and analysis. Med. Image Anal. **92**, 103046 (2024)
7. Laguna, S., et al.: Super-resolution of portable low-field MRI in real scenarios: integration with denoising and domain adaptation. In: Medical Imaging with Deep Learning (2022)
8. Koonjoo, N., Zhu, B., Bagnall, G.C., Bhutto, D., Rosen, M.S.: Boosting the signal-to-noise of low-field MRI with deep learning image reconstruction. Sci. Rep. **11**(1), 8248 (2021)
9. Shimron, E., et al.: Accelerating low-field MRI: compressed sensing and AI for fast noise-robust imaging. arXiv preprint arXiv:2411.06704 (2024)

10. Sriram, A., et al.: End-to-end variational networks for accelerated MRI reconstruction. In: Martel, A.L., et al. (eds.) MICCAI 2020. LNCS (including subseries Lecture Notes in Artificial Intelligence and Lecture Notes in Bioinformatics), vol. 12262, pp. 64–73. Springer, Cham (2020). https://doi.org/10.1007/978-3-030-59713-9_7
11. Aggarwal, H.K., Mani, M.P., Jacob, M.: MoDL: model-based deep learning architecture for inverse problems. IEEE Trans. Med. Imaging **38**(2), 394–405 (2019)
12. Zimmermann, F.F., Kofler, A.: NoSENSE: learned unrolled cardiac MRI reconstruction without explicit sensitivity maps. In: Camara, O., et al. (eds.) STACOM 2023. LNCS, vol. 14507, pp. 454–466. Springer, Cham (2023). https://doi.org/10.1007/978-3-031-52448-6_43
13. Xin, B., Ye, M., Axel, L., Metaxas, D.N.: Fill the k-space and refine the image: prompting for dynamic and multi-contrast MRI reconstruction. In: Camara, O., et al. (eds.) STACOM 2023. LNCS, vol. 14507, pp. 261–273. Springer, Cham (2023). https://doi.org/10.1007/978-3-031-52448-6_25
14. Islam, K.T., et al.: Improving portable low-field MRI image quality through image-to-image translation using paired low-and high-field images. Sci. Rep. **13**(1), 21183 (2023)
15. de Leeuw Den Bouter, M.L., Ippolito, G., O'Reilly, T.P.A., Remis, R.F., Van Gijzen, M.B., Webb, A.G.: Deep learning-based single image super-resolution for low-field MR brain images. Sci. Rep. **12**(1), 6362 (2022)
16. Tohidul Islam, K., Zhong, S., Peiris, H., Zakavi, P., Dayarathna, S., Chen, Z.: Enhancing Ultra-Low-Field MRI with Paired High-Field MRI Comparisons for Brain Imaging (ULF-EnC) (2025). https://doi.org/10.5281/zenodo.15259777
17. Loshchilov, I., Hutter, F.: Decoupled weight decay regularization. In: International Conference on Learning Representations (2019)
18. Ronneberger, O., Fischer, P., Brox, T.: U-Net: convolutional networks for biomedical image segmentation. In: Navab, N., Hornegger, J., Wells, W., Frangi, A. (eds.) MICCAI 2015. LNCS (including subseries Lecture Notes in Artificial Intelligence and Lecture Notes in Bioinformatics), vol. 9351, pp. 234–241. Springer, Cham (2015). https://doi.org/10.1007/978-3-319-24574-4_28
19. Zimmermann, F.F., et al.: MRpro - open PyTorch-based MR reconstruction and processing package (2025)
20. Wu, Y., He, K.: Group normalization. In: Ferrari, V., Hebert, M., Sminchisescu, C., Weiss, Y. (eds.) ECCV 2018. LNCS, vol. 11217, pp. 3–19. Springer, Cham (2018). https://doi.org/10.1007/978-3-030-01261-8_1
21. Perez, E., Strub, F., De Vries, H., Dumoulin, V., Courville, A.: Film: visual reasoning with a general conditioning layer. In: Proceedings of the AAAI Conference on Artificial Intelligence, vol. 32 (2018)
22. Liu, R., et al.: An intriguing failing of convolutional neural networks and the CoordConv solution. In: Advances in Neural Information Processing Systems (NeurIPS), pp. 9605–9616 (2018)
23. Goodfellow, I., et al.: Generative adversarial networks. Commun. ACM **63**(11), 139–144 (2020)
24. Dayarathna, S., et al.: McCaD: multi-contrast MRI conditioned, adaptive adversarial diffusion model for high-fidelity MRI synthesis. In: 2025 IEEE/CVF Winter Conference on Applications of Computer Vision (WACV), pp. 670–679. IEEE (2025)
25. Isola, P., Zhu, J.Y., Zhou, T., Efros, A.A.: Image-to-image translation with conditional adversarial networks. In: Proceedings of the IEEE Conference on Computer Vision and Pattern Recognition, pp. 1125–1134 (2017)

26. Miyato, T., Kataoka, T., Koyama, M., Yoshida, Y.: Spectral normalization for generative adversarial networks. In: International Conference on Learning Representations (2018)
27. Mescheder, L., Geiger, A., Nowozin, S.: Which training methods for GANs do actually converge? In: International Conference on Machine Learning, pp. 3481–3490. PMLR (2018)
28. Lehtinen, J., et al.: Noise2Noise: learning image restoration without clean data. In: International Conference on Machine Learning, pp. 2965–2974. PMLR (2018)
29. Zimmermann, F.F., Kofler, A., Kolbitsch, C., Schuenke, P.: Semi-supervised learning for spatially regularized quantitative MRI reconstruction-application to simultaneous T1 B0 B1 mapping. In: Proc. ISMRM Annu. Meeting, p. 1166 (2023)
30. Chlap, P., Min, H., Vandenberg, N., Dowling, J., Holloway, L., Haworth, A.: A review of medical image data augmentation techniques for deep learning applications. J. Med. Imaging Radiat. Oncol. **65**(5), 545–563 (2021)
31. Isensee, F., Jaeger, P.F., Kohl, S.A.A., Petersen, J., Maier-Hein, K.H.: nnU-Net: a self-configuring method for deep learning-based biomedical image segmentation. Nat. Methods **18**(2), 203–211 (2021)
32. Zhang, K., Zuo, W., Chen, Y., Meng, D., Zhang, L.: Beyond a gaussian denoiser: residual learning of deep CNN for image denoising. IEEE Trans. Image Process. **26**(7), 3142–3155 (2017)

From 64mT to 3T: Multi-sequence Low-Field MRI Enhancement via 3D U-Net

Zhicheng Lu[1(✉)], Hussain Mohammed Dipu Kabir[2], Toufique Soomro[2], and Mohammad Ali Moni[1,2]

[1] Rural Health Research Institute, Charles Sturt University, Orange, Australia
{zlu,mmoni}@csu.edu.au

[2] AI and Cyber Futures Institute, Charles Sturt University, Bathurst, Australia
{hkabir,tosoomro}@csu.edu.au

Abstract. Low-field MRI (LF-MRI), such as 64mT scanners, offers an affordable and portable solution for neuroimaging, especially in low-resource or rural settings. However, the reduced signal-to-noise ratio (SNR) and spatial resolution compared to high-field MRI (HF-MRI, 3 T) significantly limit its clinical utility. In this work, we propose a 3D U-Net-based deep learning framework to enhance multi-sequence LF-MRI to high-field MRI standards. The model takes FLAIR, T1-weighted, and T2-weighted sequences acquired from a 64mT scanner as three input channels and generates corresponding high-resolution outputs that approximate 3 T quality. By leveraging inter-sequence relationships and volumetric context, our network learns to recover high-fidelity structural and contrast details across all modalities. We train and evaluate our model using the provided paired LF-MRI and HF-MRI datasets. Preliminary results demonstrate that our method significantly improves image quality, showing potential for expanding the diagnostic capability of LF-MRI systems and bridging the performance gap between low- and high-field neuroimaging.

Keywords: Low-field MRI · Deep learning · Image enhancement

1 Introduction

Magnetic Resonance Imaging (MRI) plays a vital role in modern neuroimaging, offering rich soft-tissue contrast and multiple imaging sequences that help visualize diverse tissue properties [16]. High-field MRI (HF-MRI), typically operating at 1.5 T (T) or 3T, is widely regarded as the gold standard for clinical diagnosis due to its high spatial resolution, signal-to-noise ratio (SNR), and contrast fidelity [12,18]. Sequences such as Fluid-Attenuated Inversion Recovery (FLAIR), T1-weighted (T1W), and T2-weighted (T2W) each offer distinct advantages in detecting structural and pathological features of the brain, making them essential for stroke diagnosis, tumor detection, neurodegenerative disease monitoring, and more [10]. Despite its diagnostic advantages, HF-MRI comes with significant barriers to accessibility. The equipment is expensive, heavy,

Z. Chen et al. (Eds.): ULF-EnC 2025, LNCS 16293, pp. 37–46, 2026.
https://doi.org/10.1007/978-3-032-23344-8_5

Table 1. Configurations and estimated prices of representative medical imaging machines across different modalities including CT, MRI, and LF-MRI systems. *Note: Prices are approximate and based on publicly available vendor and reseller information as of 2025.*

Modality	Example Model	Weight	Price (USD)
CT	GE LightSpeed 16-slice	1269 kg (28–45 m^2 room)	~$135,000 [7]
MRI (1.5 T)	Siemens MAGNETOM Aera	~4800 kg (~30 m^2 room)	$150,000–$500,000 [6]
LF-MRI (64 mT)	Hyperfine Swoop	~630 kg	~$50,000 [3]
LF-MRI (0.35 T)	Neusoft Superstar	~1950 kg	~$80,000 [13]

infrastructure-dependent, and requires highly trained personnel to operate and interpret. These limitations have led to a disparity in MRI availability, particularly in rural and low-resource settings [1]. As a result, patients in these regions are less like to receive timely and effective neurological imaging, which can critically delay diagnoses and worsen outcomes, particularly for time-sensitive conditions like acute ischemic stroke.

In contrast, low-field MRI (LF-MRI), typically operating below 0.1T, offers a promising alternative to standard MRI. Portable LF-MRI devices, such as 64mT systems, are substantially more affordable, easier to maintain, and safer to deploy in rural areas [1]. Table 1 lists prices of representative medical imaging devices, it shows that the prices of LF-MRI devices are much lower than CT and standard MRI devices. Vosshenrich *et al.* reported that 0.55T MRI systems offer 40–50% cost savings in purchase compared to 1.5T systems, along with up to 70% lower installation costs and approximately 45% reduced operational expenses [19]. Furthermore, while CT and conventional HF-MRI systems typically require shielded rooms of 20–30 m^2 or more, many modern LF-MRI devices are portable, substantially lowering patient transport and infrastructure centralization costs. As a result, LF-MRI can be installed in rural clinics, emergency vehicles, or field hospitals. However, the reduced field strength of LF-MRI leads to inherent physical limitations, including poor SNR, lower spatial resolution, and limited contrast, especially in sequences critical for detecting subtle abnormalities such as early infarcts or demyelination [2]. These limitations constrain the clinical utility of LF-MRI, particularly when used as a direct replacement for HF-MRI.

To overcome these challenges, deep learning offers a transformative solution, allowing automated diagnostic workflow shown in Fig. 1(b) over that in Fig. 1(a). Image enhancement using neural networks have shown promising results in restoring degraded or under-sampled medical images. Particularly, encoder-decoder architectures such as U-Net have achieved state-of-the-art performance in medical image segmentation, synthesis, and super-resolution tasks [15]. In this work, we propose a 3D U-Net-based model LF-UNet that aims to enhance the diagnostic quality of 64mT LF-MRI images by mapping them to their 3T HF-MRI pairs. Unlike conventional approaches that treat each sequence independently, our method leverages the anatomical and fuses FLAIR, T1W, and T2W sequences as multi-channel volumetric inputs. This cross-sequence learning

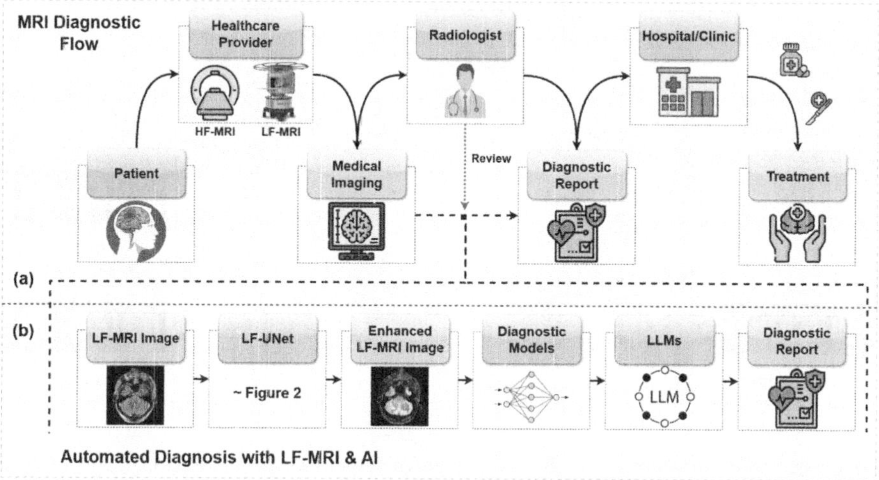

Fig. 1. (a) Traditional diagnostic workflow using HF-MRI and LF-MRI. (b) Automated diagnosis with diagnostic AI models and automated diagnosis reports with LLM. These models are to make initial diagnostic decision prior to radiologists, enabling timely diagnosis and treatment.

facilitates robust feature extraction and allows the network to recover missing contextual information more effectively. The use of a 3D architecture enables the model to fully exploit volumetric continuity and anatomical structures across slices, which is essential for preserving critical features in neuroimaging. We evaluate our method both quantitatively and qualitatively, using paired LF- and HF-MRI datasets across multiple subjects. The enhancement is assessed via peak signal-to-noise ratio (PSNR) and structural similarity (SSIM), demonstrating significant improvements over baseline methods.

In summary, our study aims to bridge the gap between accessibility and quality in neuroimaging by enhancing the diagnostic reliability of portable 64mT LF-MRI through AI-powered processing. This work has the potential to enable high-quality neuroimaging for under-resourced healthcare systems, leading to real-world clinical applications in rural and remote settings.

2 Methods

2.1 Dataset

We used the 50 paired HF-MRI and LF-MRI dataset provided by the challenge organizers for training [8,9]. Each subject scan includes co-registered FLAIR, T1W, and T2W sequences acquired using a portable Hyperfine Swoop 64mT low-field MRI system and corresponding reference images from a standard Siemens Biograph mMR 3T high-field MRI scanner using FLIRT. Bias field correction was applied using FSL-FAST, and data was converted to a standardized isotropic

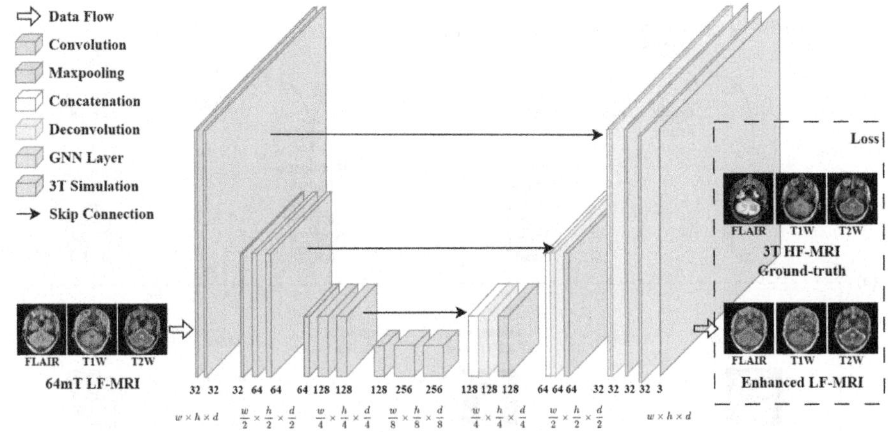

(a) Overview of the proposed LF-UNet deploys typical U-net structure, we use GNN layer (green) to replace traditional 3D-CNN layer for full extraction of 3D features.

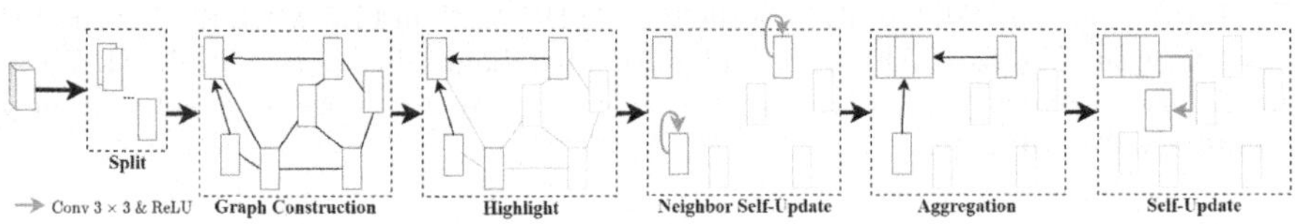

(b) Overview of GNN layer. MRI scans are split into slices and slices form nodes in the graph. Edges are constructed from k nearest neighbors of a node to the node itself. A node will intake features from its k neighbors and update itself by standard CNN filters.

Fig. 2. Details of the proposed LF-UNet.

resolution of 1 mm^3. Due to the limited size of the provided dataset, we augment by following rules: 1) crop within the range $[95\%, 100\%]$ of the original image; 2) rotate within the angle range $[-15°, 15°]$; 3) and horizontal flipping with a probability of 50%. As a result, when applying 5-fold cross validation, we have $50 \times \frac{4}{5} \times 5 = 200$ training samples.

2.2 Model Architecture

We adopted a 3D U-Net architecture as the core backbone for volumetric enhancement. The model takes a 3-channel input volume consisting of co-registered FLAIR, T1W, and T2W LF-MRI images, and outputs a 3-channel high-field target with corresponding channels. The encoder and decoder blocks consist of 3D convolutional layers with batch normalization and ReLU activations. Skip connections are used to preserve spatial details across the resolution hierarchy. The input to the LF-UNet is of size $\mathbb{R}^{3\times d\times w\times h}$, representing 3 input sequences FLAIR, T1W, and T2W from 64mT scanner. While the output is of the same size $\mathbb{R}^{3\times d\times w\times h}$, standing for the synthesized 3T sequences. To evaluate the model we performance, we capture the global structure and local intensity differences using the following equation:

$$\mathcal{L}_{total} = 0.2 \cdot \mathcal{L}1 + 0.2 \cdot \mathcal{L}2 + 0.6 \cdot \mathcal{L}_{SSIM}, \tag{1}$$

where $\mathcal{L}$ is the loss function. Flowchart of the proposed model is in Fig. 2a.

2.3 GNN Layer

Images are usually treated as grid in traditional CNN [11], where 3D features are extracted by 3D-Conv layer. However, due to the nature of CNN, extracted features from border slices might be inaccurate if we want to preserve the input size and apply padding technique. This issue is significant if border slices contains essential anatomical information. Instead, graph has more flexible structure for connected and complex objects, and based on graph structure, GNN was introduced by Scarselli *et al.* [17]. In this paper, inspired by Vision GNN proposed by Han *et al.* [5], we design our own GNN layer as illustrated in Fig. 2b. The input feature map from U-Net has size $\mathbb{R}^{c_{in} \times d \times w \times h}$, where $d \times w \times h$ is the volumetric dimensions, and c_{in} represents the number of input channels. These slices can be viewed as a set of unordered nodes in the graph as $V = \{v_1, v_2, ..., v_N\}$. For each node $v_i \in V$, its k (we set $k = 3$ in this paper) nearest neighbors $\mathcal{K}(v_i)$ will be found, and edges $e_{j,i}$ will be established from node v_j to node v_i for all $v_j \in \mathcal{K}(v_i)$. $E = \{e_{1,1}, e_{2,1}, ..., e_{k,N}\}$ denotes directed edges in the graph, by combining with nodes V we obtain graph $G = (V, E)$.

In order to extract features from each node v_i with size $v_i \in \mathbb{R}^{c_{in} \times d \times w \times h}$ and we introduce $F(\cdot)$ to update features of v_i from connected nodes:

$$\begin{aligned} v_i^{'} &= F(v_i) \\ &= \mathbf{w}_{self} \cdot (v_i \oplus \sum_{v_j \in \mathcal{K}(v_i)} \frac{\mathbf{w}_{aggr} \cdot v_j + b_1}{k}) + b_2, \end{aligned} \tag{2}$$

where $v_i^{'}$ is updated features from v_i, $\oplus$ stands for concatenation, $\mathbf{w}_{aggr}$ and $\mathbf{w}_{self}$ are learnable weights, and b_1 and b_2 are bias items. To be more specific, the kernel size of $\mathbf{w}_{aggr}$ is 3×3, together with c_{in} as both input and output channel number. The extracted feature map $\mathbb{R}^{c_{in} \times d \times w \times h}$ is then averaged and concatenated with v_i to form $\mathbb{R}^{c_{2 \cdot in} \times d \times w \times h}$ features. $\mathbf{w}_{self}$ has 3×3 kernel size, $2 \cdot c_{in}$ input channel number, and c_{out} output channel number, yielding $v_i{'} \in \mathbb{R}^{c_{out} \times d \times w \times h}$.

3 Results

3.1 Implementation Details

Model training was conducted on a workstation equipped with an Intel Core i7-13700 CPU, an NVIDIA RTX 4090 GPU, and 32 GB of RAM, running Ubuntu

20.04. The proposed models were implemented in PyTorch, with GPU acceleration enabled via CUDA 11.8. Training used the Adam optimizer with a learning rate 1×10^{-4}, a batch size of 2 using gradient accumulation was adopted due to the large size of 3D medical volumes. The network achieved convergence at approximately the 49th epoch.

Table 2. Quantitative evaluation results using 5-fold cross validation setup. SM: SSIM-Masked. PM: PSNRMasked. MM: MAEMasked. NM: NMSEMasked. SM, PM, MM, and NM are testing phase results computed on masked brain regions only (background excluded).

Methods	SSIM ↑	PSNR ↑	SM ↑	PM ↑	MM ↓	NM ↓
3D SR [4]	78.6 ± 0.6	21.7 ± 0.4	–	–	–	–
3D ResNet [14]	79.4 ± 0.7	21.5 ± 0.3	–	–	–	–
LF-UNet (no GNN)	79.8 ± 0.5	22.0 ± 0.5	–	–	–	–
LF-UNet (ours)	$\mathbf{80.5 \pm 0.4}$	$\mathbf{22.3 \pm 0.4}$	0.567	25.847	0.106	0.134

3.2 Quantitative Evaluation

To rigorously assess the performance of our proposed method, we evaluate the enhanced LF-MRI scans using two widely adopted image quality metrics: the SSIM and the PSNR. All experiments are conducted under a 5-fold cross validation setup to ensure robustness and mitigate overfitting to any specific data subset. Each fold involves training on 80% (i.e., 40) of the dataset and testing on the remaining 20% (i.e., 10), and we report the average metric values across all folds, sequences (FLAIR, T1W, and T2W), and subjects.

Table 2 summarizes the quantitative performance comparisons between our proposed LF-UNet model and several baseline methods. These include:

- **3D SRCNN** [4]: a classic 3D extension of the Super-Resolution CNN for image enhancement.
- **3D ResNet** [14]: a residual learning-based method that regresses high-field MRI from low-field input.
- **LF-UNet (without GNN)**: our ablated version of LF-UNet that excludes the GNN-based feature fusion module.

The results clearly demonstrate that LF-UNet achieves higher SSIM and PSNR scores than the baseline models, indicating better structural fidelity and reduced noise in the reconstructed high-field MRI scans. Notably, the full version of LF-UNet outperforms its ablated variant (without GNN), validating the contribution of the graph-based contextual reasoning for integrating multi-scale features across slices and sequences.

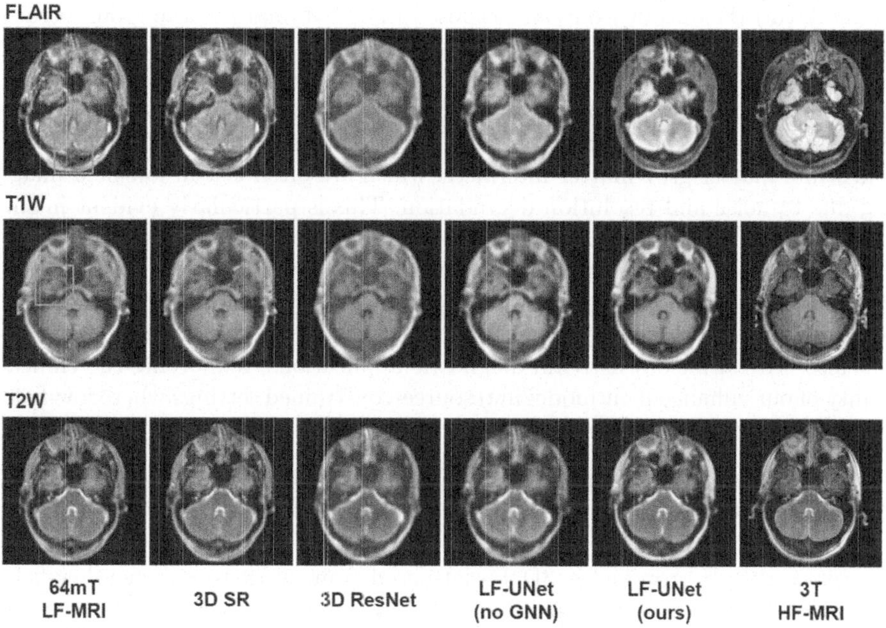

(a) Full view.

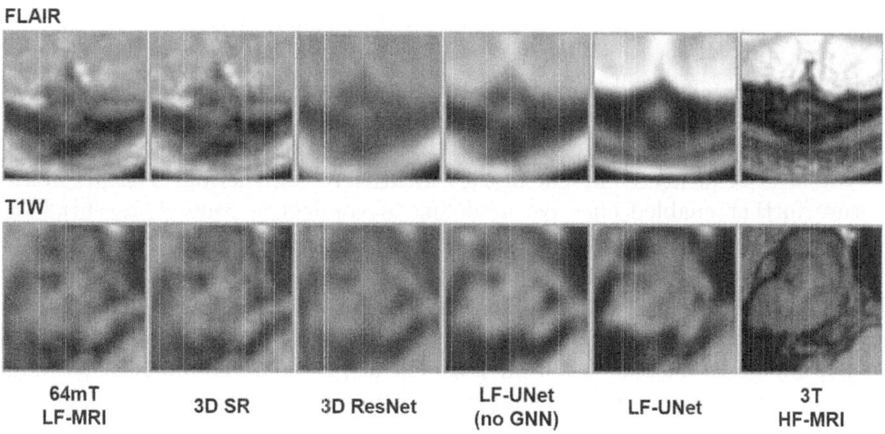

(b) Zoomed-in view of highlighted area in Figure 3a.

Fig. 3. Qualitative evaluation results on sample POCEMR001.

3.3 Qualitative Evaluation

To complement the quantitative findings, we provide visual comparisons of the reconstructed images in Fig. 3a. Each row shows a sequence, while each column shows a representative subject from the test set with four views: (i) the original input 64mT low-field MRI, (ii) the enhancement result from a selected baseline method, (iii) the output from our proposed LF-UNet, and (iv) the ground truth 3T high-field MRI scan.

Visual inspection confirms that LF-UNet produces significantly sharper and more detailed images compared to the baseline models. Our method effectively restores fine anatomical structures such as cortical gyri, ventricular boundaries, and white matter-grey matter interfaces, which are often lost in low-field imaging due to noise and resolution degradation. This is particularly evident in the FLAIR and T1W sequences, as illustrated by the zoomed-in views in Fig. 3b.

Furthermore, LF-UNet reduces background noise and enhances contrast near cerebrospinal fluid (CSF) regions, which is particularly important for stroke and lesion detection tasks. These improvements are not only aesthetically evident but also critical for downstream diagnostic applications, reinforcing the clinical utility of our enhancement model in resource-constrained settings where low-field MRI is deployed.

4 Discussion

In this work, we proposed a 3D U-Net-based framework to enhance low-field (64mT) MRI images toward high-field (3T) MRI quality, using multi-sequence input (FLAIR, T1W, T2W) as three input channels. Our model demonstrates the feasibility of leveraging deep learning to bridge the resolution and contrast gap between portable low-field systems and clinical high-field systems.

Our approach shows strong potential in improving both structural fidelity and perceptual quality of the reconstructed images, with competitive SSIM and PSNR scores. Notably, incorporating all three sequences as input channels significantly improved performance compared to single-sequence models, highlighting the value of sequence complementarity in MRI reconstruction. The 3D architecture further enabled effective modeling of volumetric context, essential for enhancing anatomical coherence across slices.

From a clinical perspective, this line of work supports the long-term goal of deploying low-cost, portable LF-MRI devices in rural and resource-constrained settings, where access to conventional HF-MRI is limited. By enabling high-quality image reconstruction from low-field acquisitions, AI models like ours may facilitate broader stroke diagnosis, neurodegenerative disease screening, and longitudinal monitoring in rural healthcare environments.

However, several limitations remain. First, while our model generalizes well on the provided dataset, further validation across diverse patient populations, scanner types, and acquisition protocols is essential. Second, improvements could include incorporating attention mechanisms, uncertainty quantification, and domain adaptation techniques to enhance robustness. Additionally, evaluating clinical usability through expert radiologist scoring or downstream diagnostic tasks (e.g., lesion detection or segmentation) will be important for real-world deployment.

In summary, this work presents a practical and effective solution for enhancing LF-MRI image quality using multi-sequence 3D deep learning. As low-field imaging continues to gain traction, such approaches could play a crucial role in advancing access to MRI worldwide.

References

1. Arnold, T.C., Freeman, C.W., Litt, B., Stein, J.M.: Low-field MRI: clinical promise and challenges. J. Magn. Reson. Imaging **57**(1), 25–44 (2023)
2. Bhat, S.S., et al.: Low-field MRI of stroke: challenges and opportunities. J. Magn. Reson. Imaging **54**(2), 372–390 (2021)
3. Deoni, S., et al.: Development of a mobile low-field MRI scanner. Sci. Rep. (2022)
4. Dong, C., Loy, C.C., He, K., Tang, X.: Image super-resolution using deep convolutional networks. IEEE Trans. Pattern Anal. Mach. Intell. **38**(2), 295–307 (2015)
5. Han, K., Wang, Y., Guo, J., Tang, Y., Wu, E.: Vision GNN: an image is worth graph of nodes. Adv. Neural. Inf. Process. Syst. **35**, 8291–8303 (2022)
6. Prizmed Imaging: 1.5t MRI machine price guide (2024). https://www.blockimaging.com/blog/1.5t-mri-machine-price-cost-guide. Accessed 03 Sept 2025
7. Prizmed Imaging: Ge lightspeed 16 CT scanner – price guide (2022). https://prizmedimaging.com/products/ge-lightspeed-16. Accessed 03 Sept 2025
8. Islam, K.T.: 2025 - ultra-low-field MRI image enhancement challenge (ULF-ENC) (2025). https://doi.org/10.7303/SYN65485242. https://repo-prod.prod.sagebase.org/repo/v1/doi/locate?portalId=1&id=syn65485242&type=ENTITY
9. Islam, K.T., et al.: Improving portable low-field MRI image quality through image-to-image translation using paired low-and high-field images. Sci. Rep. **13**(1), 21183 (2023)
10. Lavrova, A., et al.: Quality assessment of routine brain imaging at 0.55 t: initial experience in a clinical workflow. NMR Biomed. **37**(7), e5017 (2024)
11. Lu, Z., Yeung, H.W., Qu, Q., Chung, Y.Y., Chen, X., Chen, Z.: Improved image classification with 4d light-field and interleaved convolutional neural network. Multimedia Tools Appl. **78**, 29211–29227 (2019)
12. Mathivanan, S.K., Sonaimuthu, S., Murugesan, S., Rajadurai, H., Shivahare, B.D., Shah, M.A.: Employing deep learning and transfer learning for accurate brain tumor detection. Sci. Rep. **14**(1), 7232 (2024)
13. Medical Price Online: Neusoft superstar 0.35t (2025). https://www.medicalpriceonline.com/medical-equipment/product/neusoft-superstar-0-35t. Accessed 03 Sept 2025
14. Qiu, Z., Yao, T., Mei, T.: Learning spatio-temporal representation with pseudo-3d residual networks. In: Proceedings of the IEEE International Conference on Computer Vision, pp. 5533–5541 (2017)
15. Ronneberger, O., Fischer, P., Brox, T.: U-Net: convolutional networks for biomedical image segmentation. In: Navab, N., Hornegger, J., Wells, W., Frangi, A. (eds.) MICCAI 2015. LNCS, vol. 9351, pp. 234–241. Springer, Cham (2015). https://doi.org/10.1007/978-3-319-24574-4_28
16. Scalas, G., et al.: Magnetic resonance imaging of soft tissue sarcoma: features related to prognosis. Eur. J. Orthopaedic Surg. Traumatol. **31**(8), 1567–1575 (2021)

17. Scarselli, F., Gori, M., Tsoi, A.C., Hagenbuchner, M., Monfardini, G.: The graph neural network model. IEEE Trans. Neural Networks **20**(1), 61–80 (2008)
18. Stewart, N.J., et al.: Lung MRI with hyperpolarised gases: current & future clinical perspectives. Br. J. Radiol. **95**(1132), 20210207 (2022)
19. Vosshenrich, J., Breit, H.C., Bach, M., Merkle, E.M.: Economic aspects of low-field magnetic resonance imaging: acquisition, installation, and maintenance costs of 0.55 t systems. Der Radiologe **62**(5), 400–404 (2022)

Multi-view Fusion-Guided Brownian Bridge Diffusion Model for Ultra-Low-Field MRI Enhancement

Mengfei Wang[1,2], Ruipeng Zhang[2], Yidong Jin[2], and Yuehua Li[1,2](✉)

[1] DongHua University, Shanghai, China
liyuehua77@sjtu.edu.cn
[2] Institute of Diagnostic and Interventional Radiology, Shanghai Sixth People's Hospital Affiliated to Shanghai Jiao Tong University School of Medicine, Shanghai, China
zhangrp@sjtu.edu.cn

Abstract. Ultra-low-field (ULF) MRI promises to make diagnostic imaging more accessible, yet its clinical adoption is impeded by poor image quality. Enhancing ULF MRI is confronted by a dual challenge: the immense computational complexity of direct 3D models and the critical scarcity of paired training data. In this paper, we address this by drawing inspiration directly from clinical practice and propose the Multi-view Fusion-guided Brownian Bridge Diffusion Model (MF-BBDM), a framework that computationally emulates how radiologists fluidly navigate between axial, sagittal, and coronal views to build a comprehensive 3D anatomical understanding. By decomposing the intractable 3D task into a set of efficient 2D translations and subsequently fusing their results, our framework is both data-efficient and ensures global anatomical consistency. On the official MICCAI 2025 ULF-EnC validation set, our MF-BBDM demonstrated strong quantitative performance, achieving a PSNR of 19.529 dB and an SSIM of 0.733. This result validates our clinically-inspired approach, demonstrating its ability to produce images of true diagnostic quality and paving the way for the widespread adoption of accessible ULF MRI.

Keywords: Ultra-low-field MRI · Diffusion-based Image Enhancement · Multi-view Fusion

1 Introduction

While Magnetic Resonance Imaging (MRI) is a vital diagnostic tool, renowned for its superior soft-tissue contrast and safety without ionizing radiation [14,19], conventional high-field systems are hindered by high costs, large infrastructure demands, and patient contraindications, limiting their widespread accessibility [20,26]. To address these barriers, ultra-low-field (ULF) MRI has emerged as a promising solution, offering portability, reduced cost, and deployment flexibility

Z. Chen et al. (Eds.): ULF-EnC 2025, LNCS 16293, pp. 47–56, 2026.
https://doi.org/10.1007/978-3-032-23344-8_6

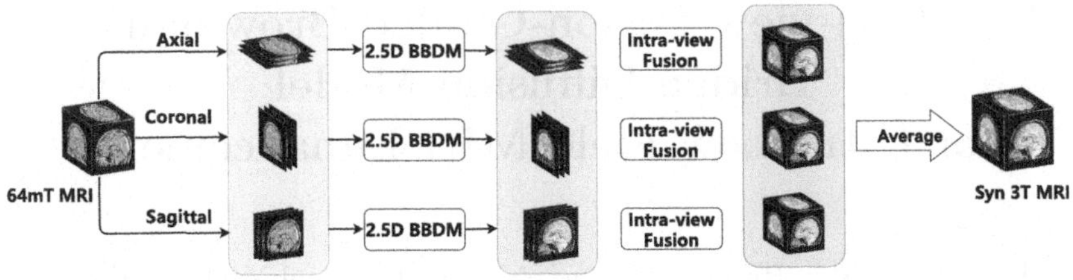

Fig. 1. The overall pipeline of our proposed Multi-view Fusion-guided Brownian Bridge Diffusion Model (MF-BBDM). A 3D ULF MRI image is first decomposed into three orthogonal views. Each view is processed by an independent BBDM using a 2.5D slice-stacking strategy. The multi-slice outputs from each model then undergo an intra-view fusion to reconstruct a coherent 3D volume for that view. Finally, the three resulting volumes are averaged to produce the final high-quality 3D image.

to extend imaging access to new settings [4,11,25]. Yet, this immense potential is fundamentally constrained by a critical trade-off inherent to its design: the extremely low field strength (e.g., 64mT) results in acquisitions with a poor signal-to-noise ratio (SNR) and low spatial resolution. This performance gap severely limits its diagnostic utility compared to mainstream high-field MRI [1], making the development of effective image enhancement algorithms a critical and pressing research challenge to unlock its full clinical value.

In recent years, deep learning-based methods have achieved remarkable success in medical image enhancement tasks [2,5,7,9,13,16,22]. Generative models, like Generative Adversarial Networks [8] (GANs) and Denoising Diffusion Probabilistic Models [10] (DDPM), have shown great promise for improving image quality, thanks to their powerful capacity for modeling complex data distributions [3,6,17,18]. However, enhancing low-quality ULF MRI is confronted by a dual challenge: the immense computational complexity of direct 3D modeling and, critically, the scarcity of large-scale paired datasets required to train such data-hungry models. Our primary motivation, therefore, is drawn directly from clinical practice. We observed that radiologists do not perceive anatomy from a single perspective; instead, they mentally reconstruct a patient's 3D anatomy by fluidly navigating between axial, sagittal, and coronal views, using information from each plane to resolve ambiguities and build a comprehensive understanding.

Inspired by this expert diagnostic workflow, we propose a novel **Multi-view Fusion-guided Brownian Bridge Diffusion Model (MF-BBDM)**. The core design insight is to computationally emulate this multi-view reasoning process. By decomposing the intractable 3D enhancement task into three more manageable 2D translations, our framework not only becomes highly data-efficient—critically maximizing the utility of limited paired volumes by treating each 2D slice as a distinct training sample—but also inherently robust. The final fusion of these orthogonal views ensures global anatomical consistency, integrating complementary contextual information for a more detailed and reliable reconstruction, much like a radiologist synthesizes information from all views.

Operationally, we train independent Brownian Bridge Diffusion Models (BBDM) [15] on 2.5D inputs [21,23,24] and employ a sophisticated multi-stage fusion process during inference. Rigorously evaluated on the MICCAI 2025 Ultra-Low-Field MRI Enhancement Challenge (ULF-EnC) dataset [12], our MF-BBDM achieved state-of-the-art performance on metrics like PSNR and SSIM. This result provides the ultimate validation for our approach, demonstrating that by mimicking the expert's diagnostic process, our model can produce results that satisfy the expert's eye. The main contributions of this paper are as follows:

- We propose MF-BBDM, a novel framework inspired by radiological practice that decomposes the intractable 3D enhancement task into a set of efficient 2D translations on orthogonal planes.
- We design a dedicated multi-stage fusion pipeline that integrates the enhanced orthogonal views to reconstruct globally coherent and high-quality 3D volumes.
- We demonstrate competitive performance on a public benchmark dataset through extensive quantitative evaluation, showing that our method effectively recovers high-field MRI quality from low-field input.

2 Method

Our proposed framework, MF-BBDM, enhances 3D ULF MRI by decomposing the problem into a series of 2D image-to-image translations across three orthogonal views, followed by a multi-stage fusion process to ensure 3D consistency. The overall architecture of our proposed MF-BBDM is illustrated in Fig. 1.

2.1 Dataset and Preprocessing

Dataset. We utilized the official dataset from the MICCAI 2025 Ultra-Low-Field MRI Enhancement Challenge (ULF-EnC) [12]. As per the challenge description, the dataset contains paired brain MRI scans from a 64mT system (Hyperfine Swoop) and a 3T system (Siemens Biograph mMR). The data includes T1-weighted, T2-weighted, and FLAIR contrasts for each subject. The dataset is officially partitioned into 50 cases for training, 10 for validation, and 15 for testing. For this study, the 3T scans serve as the ground truth reference for their paired 64mT counterparts.

Official Preprocessing. The challenge organizers performed a standardized preprocessing pipeline on all images, which included:

1. Conversion from DICOM to NIFTI format.
2. Resampling to a standardized isotropic resolution of $1 \times 1 \times 1$ mm^3.
3. Bias field correction using the FSL-FAST tool [27].
4. Co-registration of 0.064T and 3T scans for each subject using FSL-FLIRT to ensure precise spatial alignment.

Our Preprocessing for Model Input. Building on the pre-aligned data, we prepared the inputs for our model as follows:

1. **2.5D Slice Stacking:** To provide the 2D network with local 3D context, we adopted a 2.5D strategy. For each anatomical plane, we treated each 2D slice as a three-channel image by stacking it with its immediate neighbors. Specifically, to generate the enhanced version of slice i, the model input is a stack of $[\text{slice}_{i-1}, \text{slice}_i, \text{slice}_{i+1}]$. For boundary slices (i.e., the first and last slices of a volume), missing neighbors were padded with zeros.
2. **Intensity Normalization:** The intensity values of all input slices were normalized to the range of $[-1, 1]$ using min-max normalization to ensure stable network training. No further cropping or resampling was performed.

2.2 Multi-view Brownian Bridge Diffusion Model

Our core enhancement engine consists of three independent Brownian Bridge Diffusion Models (BBDM), each dedicated to one of the three orthogonal views.

Brownian Bridge Formulation. For an image-to-image translation task like enhancing ULF MRI to 3T quality, the goal is not to generate an image from pure noise, but rather to transform a given source image (x_0, the ULF MRI) into a target image (x_T, the 3T MRI). We chose the Brownian Bridge Diffusion Model (BBDM) over a standard Denoising Diffusion Probabilistic Model (DDPM) for its theoretical suitability for this task. A standard DDPM learns the reverse process of converting noise x_T into a target image, conditioned on a source image. In contrast, a BBDM models a stochastic process that is anchored by **both** the source image x_0 at time $t = 0$ and the target image x_T at time $t = T$. The forward noising process is a "bridge" directly connecting the two domains. Its probability path is defined as:

$$p(x_t|x_0, x_T) = \mathcal{N}\left(x_t; \left(1 - \frac{t}{T}\right) x_0 + \frac{t}{T} x_T, \frac{t(T-t)}{T}\mathbf{I}\right) \tag{1}$$

This formulation offers two key advantages: **More Direct Transformation Path:** The learning objective is to reverse this bridging process, which is a more constrained and direct mapping from the source distribution to the target distribution compared to the conditional generation in standard DDPMs. This encourages the preservation of structural information from the source image. **Improved Sampling Efficiency:** The variance of the bridge process, $\frac{t(T-t)}{T}$, is smaller than that of a standard forward process. This results in a "shorter" path between the domains, which can potentially lead to higher quality results with fewer sampling steps during inference.

Implementation Details. The reverse process of our BBDM is parameterized by a U-Net-based neural network, ϵ_θ. The network takes a 6-channel input (concatenation of the 3-channel 2.5D condition and the 3-channel noisy image) and

Table 1. Architectural and training hyperparameters of our MF-BBDM.

Parameter Category	Value/Specification
Network Architecture (U-Net)	
Input Size/Channels	$224 \times 224/6$ channels
Initial Feature Channels	128
Channel Multipliers	(1, 2, 4)
Residual Blocks per Level	2
Attention Resolutions	128×128, 64×64, 32×32
Attention Heads	8
Normalization	Scale-shift normalization
Diffusion Process	
Total Timesteps (T)	1000
Noise Schedule	Linear
Inference Sampler	DDIM-like ($\eta = 1.0$)
Inference Steps	50
Training Details	
Optimizer	Adam ($\beta_1 = 0.9$, no weight decay)
Learning Rate (Initial)	1×10^{-4}
LR Scheduler	Reduce on plateau (factor = 0.5, patience = 3k steps)
Batch Size	8
Total Training Steps	50,000 (20 epochs)
EMA Decay Rate	0.995

is trained to predict the added noise. We optimized the network using a Mean Absolute Error (L1) loss, which encourages the generation of sharper images compared to an L2 loss:

$$\mathcal{L}_{L1} = \mathbb{E}_{t,x_0,x_T,\epsilon}\left[||\epsilon - \epsilon_\theta(x_t, t, x_0)||\right] \tag{2}$$

The detailed architectural and training hyperparameters are summarized in Table 1. All models were implemented in PyTorch 2.2 and trained on a single NVIDIA A100 GPU.

2.3 Multi-stage Fusion for 3D Reconstruction

A final, high-quality 3D volume is reconstructed by the two-stage fusion process that leverages the complementary information from three orthogonal views.

Stage 1: Intra-view Reconstruction and Fusion. For each view (axial, sagittal, coronal), the trained BBDM processes all 2.5D input stacks. As previously described, the model outputs three corresponding enhanced slices for each

Algorithm 1. Edge-Preserving Intra-view Fusion

Require: A list of N consecutive enhanced slice stacks, $L = \{E_1, E_2, ..., E_N\}$. (In our case, $N = 3$). Parameters: $\sigma_{spatial}, \sigma_{intensity}, \tau_{edge}$ (edge threshold).

Ensure: A single fused slice, F_{final}.

```
procedure EdgePreservingFusion(L, σ_spatial, σ_intensity, τ_edge)
    Initialize F_fused ← zeros_like(E_1).
    for each enhanced slice stack E_k in L do
        S_k ← BilateralFilter(E_k, σ_spatial, σ_intensity)    ▷ Generate smoothed component
        M_k ← DetectEdges(E_k, τ_edge)    ▷ Generate edge mask
        w_k ← 0.2 + 0.8 × (sum(M_k)/size(M_k))    ▷ Calculate weight based on edge density
        A_k ← where(M_k, E_k, S_k)    ▷ Adaptively recombine using the edge mask
        F_fused ← F_fused + w_k × A_k    ▷ Accumulate weighted images
    end for
    F_final ← F_fused/N    ▷ Normalize the result by the number of stacks
    return F_final
end procedure

function BilateralFilter(V, σ_s, σ_i)
    return Apply slice-wise 2D bilateral filter to volume V.
end function

function DetectEdges(V, τ_e)
    G_mag ← GradientMagnitude(V)    ▷ Compute 3D gradient magnitude
    return G_mag > (τ_e × std(G_mag))    ▷ Return binary mask based on adaptive threshold
end function
```

input position i, denoted as $\{e_{i-1}, e_i, e_{i+1}\}$. To merge this contextual output into a single, structurally coherent 2D slice, we designed a custom edge-preserving weighted fusion algorithm. This algorithm operates as follows:

1. **Component Generation:** For each of the three output slices, we first generate two components. A *base component* is created by applying a 2D bilateral filter slice-by-slice to smooth the image while preserving major edges. An *edge mask* is simultaneously generated by computing the 3D gradient magnitude across the three stacked slices ($\sqrt{g_x^2 + g_y^2 + g_z^2}$) and applying an adaptive threshold proportional to the gradient's standard deviation.
2. **Adaptive Image Recombination:** An "adaptive image" is created for each of the three slices. This is achieved by taking pixel values from the original enhanced slice in regions identified as edges by the mask, and from the bilaterally filtered base component in non-edge (smooth) regions. This step selectively enhances structural sharpness.
3. **Weighted Averaging:** The final fused slice, f_i, is produced by a weighted average of the three adaptive images. The weight for each adaptive image is calculated based on its structural complexity, specifically, the proportion of

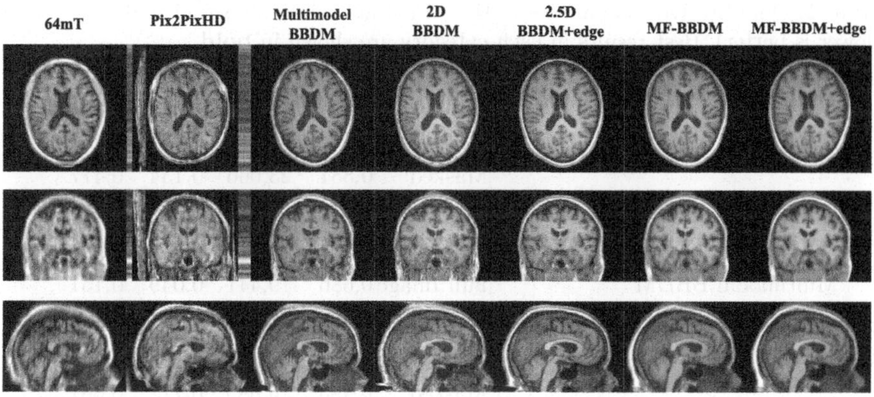

Fig. 2. Qualitative comparison on a representative case in three orthogonal views. Diffusion-based models progressively improve upon the artifact-ridden GAN result. Our proposed **MF-BBDM** yields the highest structural fidelity, best preserving fine anatomical details like the cortical gray-white matter boundary and sulcal patterns.

edge pixels within it. Images with more detected edges are assigned a higher weight, contributing more to the final result.

This procedure is repeated for all slice positions, and the resulting fused 2D slices $\{f_i\}$ are re-stacked to form a complete, enhanced 3D volume for that specific view. This process yields three distinct 3D volumes: V_{axial}, V_{sagittal}, and V_{coronal} (Fig. 2).

Stage 2: Inter-view Fusion. To create the final output volume, we fused the three intermediate volumes from the previous stage using a voxel-wise averaging:

$$V_{\text{final}} = \frac{V_{\text{axial}} + V_{\text{sagittal}} + V_{\text{coronal}}}{3} \tag{3}$$

This simple yet effective method reduces view-specific artifacts and enhances the overall structural consistency and signal-to-noise ratio of the final 3D image.

3 Experimental Results

Quantitative Results. Performance on the official validation set is summarised in Table 2. To clearly present all eight official metrics, the table groups the results for each method into 'Full Image' and 'Masked' (intracranial region) categories. The proposed MF-BBDM achieves the best or joint-best scores across the board, with particularly notable gains in the structurally-sensitive metrics such as SSIM (0.733) and maskSSIM (0.514).

Table 2. Quantitative comparison on the ULF-EnC validation set (↑ higher is better, ↓ lower is better). Best results in each category are shown in **bold**.

Method	Scope	SSIM ↑	PSNR ↑	MAE ↓	MSE ↓
Pix2PixHD	Full Image	0.410	15.112	0.104	0.421
	Masked	0.381	23.066	0.134	0.477
2D BBDM	Full Image	0.707	19.379	0.050	0.151
	Masked	0.465	25.819	0.107	0.337
Multimodal BBDM	Full Image	0.686	19.441	0.049	0.151
	Masked	0.484	25.973	0.105	0.339
2.5D BBDM w/ Edge-Preserving	Full Image	0.727	19.128	0.050	0.163
	Masked	0.495	26.085	0.112	0.381
MF-BBDM + edge	Full Image	0.727	19.521	0.050	0.149
	Masked	0.513	25.796	0.109	0.350
MF-BBDM (Ours)	Full Image	**0.733**	**19.529**	**0.049**	**0.149**
	Masked	**0.514**	**25.811**	**0.109**	**0.350**

Discussion. Three key observations emerge from our results (Table 2). First, **diffusion models substantially outperform the GAN baseline** (Pix2PixHD) in both full-image and masked evaluations, underscoring their strong generative capacity for this task. Second, our **ablation study validates the contribution of each proposed component**. Moving from a slice-wise 2D BBDM to a **2.5D BBDM with intra-view fusion** yields clear improvements in structural metrics, particularly in the masked SSIM (0.465 to 0.495). Subsequently, introducing *multi-view fusion* (our final MF-BBDM) further boosts performance across the board, demonstrating that integrating complementary information across views effectively mitigates slice-wise artefacts and enforces global 3D consistency. Finally, the ablation on the *inter-view fusion strategy* ('MF-BBDM + edge') reveals that replacing the final averaging step with a second, more complex fusion stage does not yield further improvements, indicating that simple averaging is a more efficient and equally effective choice.

In summary, the strong performance of MF-BBDM stems from its carefully designed pipeline that progressively exploits 2.5D context, edge-preserving intra-view fusion, and lightweight inter-view averaging. The pronounced advantage on masked metrics, which directly assess the quality of the reconstructed intracranial structures, highlights its suitability for downstream neuroimaging applications where anatomical accuracy is critical.

4 Discussion

It is well-understood that deep learning models, especially generative ones like BBDM, are data-hungry. Our 2.5D approach, while computationally efficient, inherently requires the model to learn robust representations from a wider variety of 2D anatomical views. Our MF-BBDM's architectural choice offers a key advantage over FCNN-based AE and GANs. Its flexible, probabilistic translation surpasses their rigid, deterministic mappings, yielding competitive results even with limited data. Crucially, this positions our framework to uniquely benefit from future data scaling, ensuring a more promising trajectory for performance growth and clinical adoption. In conclusion, this work presents a practical and effective pathway for enhancing low-SNR 3D medical images. Future efforts will be directed towards applying the framework to clinical data featuring various pathologies and exploring methods for inference acceleration to further support the widespread use of low-cost, portable ULF-MRI.

Acknowledge. This work was supported by the National Natural Science Foundation of China (Grant No. 8225024), the Key R&D Sub-project of the Ministry of Science and Technology (Grant No. 2023YFF1204804), the Shanghai Municipal Commission of Science and Technology Explorer Program (Grant No. 23TS1400400), the Medicine and Engineering Interdisciplinary Program of Shanghai Jiao Tong University (Grant No. YG2024LC08), and the Shanghai Key Discipline of Medical Imaging (Grant No. 2017ZZ02005). The authors gratefully acknowledge the financial support and resources provided by these institutions, which have been instrumental in advancing this research.

References

1. Arnold, T.C., Freeman, C.W., Litt, B., Stein, J.M.: Low-field MRI: clinical promise and challenges. J. Magn. Reson. Imaging **57**(1), 25–44 (2023)
2. Chen, Z., Pawar, K., Ekanayake, M., Pain, C., Zhong, S., Egan, G.F.: Deep learning for image enhancement and correction in magnetic resonance imaging–state-of-the-art and challenges. J. Digit. Imaging **36**(1), 204–230 (2023)
3. Choo, K., Jun, Y., Yun, M., Hwang, S.J.: Slice-consistent 3d volumetric brain CT-to-MRI translation with 2D Brownian bridge diffusion model. In: International Conference on Medical Image Computing and Computer-Assisted Intervention, pp. 657–667. Springer (2024)
4. Cooley, C.Z., et al.: Design and implementation of a low-cost, tabletop MRI scanner for education and research prototyping. J. Magn. Reson. **310**, 106625 (2020)
5. Dayarathna, S., Islam, K.T., Chen, Z.: Ultra low-field to high-field MRI translation using adversarial diffusion. In: 2024 IEEE International Symposium on Biomedical Imaging (ISBI), pp. 1–4. IEEE (2024)
6. Dayarathna, S., Islam, K.T., Uribe, S., Yang, G., Hayat, M., Chen, Z.: Deep learning based synthesis of MRI, CT and PET: review and analysis. Med. Image Anal. **92**, 103046 (2024)
7. Dayarathna, S., et al.: MCCAD: multi-contrast MRI conditioned, adaptive adversarial diffusion model for high-fidelity MRI synthesis. In: 2025 IEEE/CVF Winter Conference on Applications of Computer Vision (WACV), pp. 670–679. IEEE (2025)
8. Goodfellow, I.J., et al.: Generative adversarial nets. Adv. Neural Inf. Process. Syst. **27** (2014)
9. He, C., et al.: HQG-Net: unpaired medical image enhancement with high-quality guidance. IEEE Trans. Neural Netw. Learn. Syst. (2023)

10. Ho, J., Jain, A., Abbeel, P.: Denoising diffusion probabilistic models. Adv. Neural. Inf. Process. Syst. **33**, 6840–6851 (2020)
11. Hori, M., Hagiwara, A., Goto, M., Wada, A., Aoki, S.: Low-field magnetic resonance imaging: its history and renaissance. Invest. Radiol. **56**(11), 669–679 (2021)
12. Islam, K.T.: 2025 - ultra-low-field MRI image enhancement challenge (ULF-ENC) (2025). https://doi.org/10.7303/SYN65485242. https://repo-prod.prod.sagebase.org/repo/v1/doi/locate?portalId=1&id=syn65485242&type=ENTITY
13. Islam, K.T., et al.: Improving portable low-field MRI image quality through image-to-image translation using paired low-and high-field images. Sci. Rep. **13**(1), 21183 (2023)
14. Lauterbur, P.C.: Image formation by induced local interactions: examples employing nuclear magnetic resonance. Nature **242**(5394), 190–191 (1973)
15. Li, B., Xue, K., Liu, B., Lai, Y.K.: BBDM: image-to-image translation with Brownian bridge diffusion models. In: Proceedings of the IEEE/CVF Conference on Computer Vision and Pattern Recognition, pp. 1952–1961 (2023)
16. Li, H., et al.: Enhancing and adapting in the clinic: source-free unsupervised domain adaptation for medical image enhancement. IEEE Trans. Med. Imaging **43**(4), 1323–1336 (2023)
17. Longuefosse, A., et al.: Generating high-resolution synthetic CT from lung MRI with ultrashort echo times: initial evaluation in cystic fibrosis. Radiology **308**(1), e230052 (2023)
18. Lyu, J., et al.: Generative adversarial network-based noncontrast CT angiography for aorta and carotid arteries. Radiology **309**(2), e230681 (2023)
19. Mansfield, P., Maudsley, A.A.: Medical imaging by NMR. Br. J. Radiol. **50**(591), 188–194 (1977)
20. Expert Panel on MR Safety, E.P., et al.: ACR guidance document on MR safe practices: 2013. J. Magn. Reson. Imaging **37**(3), 501–530 (2013)
21. Nie, D., et al.: Medical image synthesis with context-aware generative adversarial networks. In: International Conference on Medical Image Computing and Computer-Assisted Intervention, pp. 417–425. Springer (2017)
22. Oulmalme, C., Nakouri, H., Jaafar, F.: A systematic review of generative AI approaches for medical image enhancement: comparing GANs, transformers, and diffusion models. Int. J. Med. Inform., 105903 (2025)
23. Prasoon, A., Petersen, K., Igel, C., Lauze, F., Dam, E., Nielsen, M.: Deep feature learning for knee cartilage segmentation using a triplanar convolutional neural network. In: International Conference on Medical Image Computing and Computer-Assisted Intervention, pp. 246–253. Springer (2013)
24. Roth, H.R., et al.: A new 2.5 d representation for lymph node detection using random sets of deep convolutional neural network observations. In: International Conference on Medical Image Computing and Computer-Assisted Intervention, pp. 520–527. Springer (2014)
25. Sarracanie, M., LaPierre, C.D., Salameh, N., Waddington, D.E., Witzel, T., Rosen, M.S.: Low-cost high-performance MRI. Sci. Rep. **5**(1), 15177 (2015)
26. Westbrook, C., Talbot, J.: MRI in Practice. Wiley (2018)
27. Zhang, Y., Brady, M., Smith, S.: Segmentation of brain MR images through a hidden Markov random field model and the expectation-maximization algorithm. IEEE Trans. Med. Imaging **20**(1), 45–57 (2002)

Bridged Denoising Diffusion in Ultra-Low Field MRI Enhancement Challenge

Baris Imre[1(✉)], Chinmay Rao[1], Aram Salehi[2,3], Marius Staring[1], and Efe Ilicak[1]

[1] Division of Image Processing, Department of Radiology, Leiden University Medical Center, Leiden, The Netherlands
b.imre@lumc.nl

[2] Department of Human Genetics, Radboud University Medical Center, Nijmegen, The Netherlands

[3] Department of Radiology, Leiden University Medical Center, Leiden, The Netherlands

Abstract. Low-field MRI offers a cost-effective and portable alternative to conventional high-field systems, but its clinical use is limited by reduced image quality. Recent advancements in deep learning have aimed to enhance low-field scans, including approaches based on GANs, stochastic quality transfer, and denoising diffusion models. In this paper, we introduce a bridged denoising diffusion model that explicitly aligns the latent spaces of low- and high-field MR images through a learned bridging model. At a predefined timestep in the diffusion process, a bridge network translates the noisy low-field representation into the high-field domain, enabling the downstream diffusion model, trained solely on high-field data, to generate high-field-like images. The proposed method is trained on paired 3T and 0.64 mT scans across multiple contrasts, based on the "Enhancing Ultra-Low-Field MRI with Paired High-Field MRI Comparisons for Brain Imaging" challenge. On the hidden validation set, our method achieved an SSIM of 0.779 and a PSNR of 22.85 dB, outperforming alternative configurations in ablation experiments. We show that our model can significantly enhance low-field images.

Keywords: Low-field enhancement · MRI enhancement · Denoising diffusion

1 Introduction

MRI is a cornerstone imaging modality due to its excellent soft tissue contrast and non-ionizing nature [1]. It offers versatile imaging capabilities across multiple anatomical regions and clinical indications, while being an extremely valuable research tool. Despite its advantages, conventional MRI systems are often expensive, immobile, and require significant operational infrastructure. The limited accessibility of conventional MR scanners is primarily due to their high costs. [2,3] (Fig. 1).

To address these, low-field (<0.1 T) MR scanners have recently seen a resurgence in interest as a more accessible and cost-effective alternative [4]. With

Z. Chen et al. (Eds.): ULF-EnC 2025, LNCS 16293, pp. 57–66, 2026.
https://doi.org/10.1007/978-3-032-23344-8_7

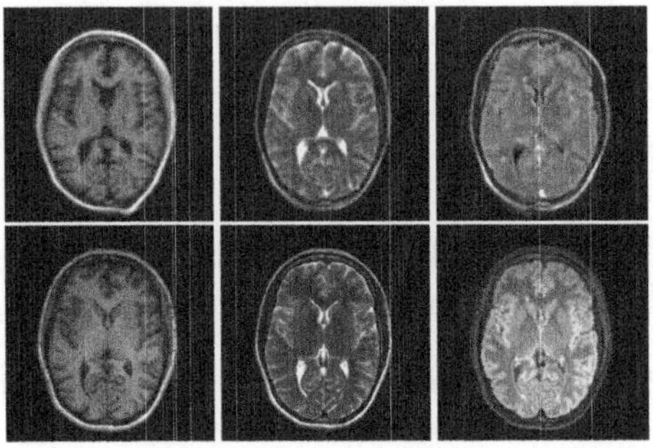

Fig. 1. A comparison of paired 3T and 0.64 mT MR images in three different contrasts. From left to right: T1-w, T2-w, FLAIR. Top row: low-field, bottom row: high-field.

portable designs based on permanent magnets and lower infrastructural requirements, these devices can be deployed outside conventional MRI suites, such as for point-of-care imaging [5,6]. While recent works have demonstrated the feasibility of these systems in detecting numerous conditions [7,8], obtaining diagnostic-quality images with these systems remains a challenge [9].

Approaches to low-field MRI enhancement have included image-to-image translation methods based on GANs, which aim to synthesize high-field quality images from paired low-field inputs [10]. While effective, such models can suffer from instability and hallucinated features. Other efforts have explored alternative strategies such as stochastic image quality transfer, which avoids adversarial training by learning uncertainty-aware mappings between low and high-field images [11]. In parallel, denoising diffusion models have gained momentum in the MRI community, particularly in the context of accelerated MRI reconstruction [12,13]. Although originally applied to reconstruct undersampled k-space acquisitions, the same modeling principles have been adapted to the low-field enhancement setting with promising results as well [14,15].

In this paper, we propose a bridged denoising diffusion model and training scheme to enhance low-field images to their paired high-field counterparts. We describe our methodology in depth and present results on a challenge setting.

2 Materials and Methods

Figure 2 shows an overview of the bridged denoising diffusion model that we propose. Our method has two main components, namely the bridge and the denoising diffusion. In both of the components, we combine the three contrasts of each sample along the channel dimension, promoting the networks to learn and cross reference using all three contrasts.

2.1 Dataset and Task

We conduct all of our experiments with a dataset of 50 paired 3D MR images with an addition of a validation and test set that is withheld from us. All

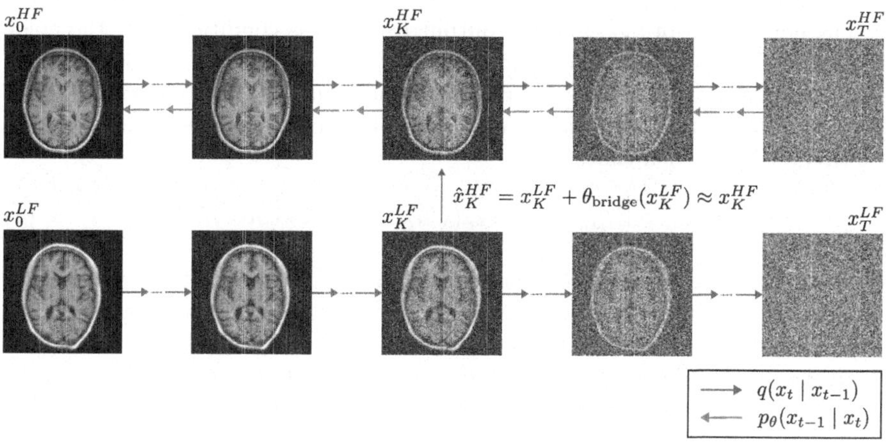

Fig. 2. An overview of the bridged denoising diffusion model for low-field MR enhancement. Two diffusion tracks are presented. On top: high-field, below: low-field. The timesteps from left to right range from $t = 0$ to $t = T$. The bridging step, parameterized by the bridge network ϕ_{bridge} bridges the gap between x_{LF} and x_{HF} at $t = K$.

scans are resampled to an isotropic resolution of $1\,\text{mm}^3$. The pairs consist of 3T and 0.64mT MR scans in T1-w, T2-w, and FLAIR contrasts. All scans are co-registered. The challenge data is acquired using a Siemens Biograph mMR (3T) scanner and a Hyperfine Swoop (64mT) scanner. An example is presented in Sect. 1. We measure the average SSIM [16] index and the PSNR of the paired samples as 0.547 and 17.35 dB respectively. The enhancement task is therefore defined as producing pseudo high-field samples that score higher.

2.2 Denoising Diffusion

Denoising diffusion models are defined with the forward operator as defined by Ho et al. [17]:

$$q(x_t \mid x_0) = \mathcal{N}(x_t; \sqrt{\bar{\alpha}_t}x_0, (1 - \bar{\alpha}_t)\mathbf{I}) \tag{1}$$

Given the clean starting sample x_0, we derive a noisy version x_t at timestep t with the noise coefficient $\bar{\alpha}_t$ where $\bar{\alpha}_t = \prod_{s=1}^{t} \alpha_s$ and $\alpha_s = 1 - \beta_s$ and β_s is a predefined noise schedule. This results in Gaussian noise being added on top of the clean sample gradually. We use a cosine based noise schedule. The objective of the denoising diffusion network is to reverse this process by predicting the noise at the given timestep and denoise with the following process:

$$p_\theta(x_{t-1} \mid x_t) = \mathcal{N}\left(x_{t-1}; \mu_\theta(x_t, t), \sigma_t^2\mathbf{I}\right) \tag{2}$$

defining,

$$\mu_\theta(x_t, t) = \frac{1}{\sqrt{\alpha_t}}\left(x_t - \frac{\beta_t}{\sqrt{1 - \bar{\alpha}_t}}\epsilon_\theta(x_t, t)\right) \tag{3}$$

The neural network ϵ_θ, with its output $\epsilon_\theta(x_t, t)$, is conditioned by the noisy image and the timestep. The training objective is defined as the negative log likelihood of the Gaussian noise prediction and sample, which reduces to an ℓ_2 norm for Gaussian distributions. We define this training objective, hence loss function as $\mathcal{L}_{\text{noise}}$ such that:

$$\mathcal{L}_{\text{noise}} = \mathbb{E}_{t,x_0,\epsilon}\left[\left|\left|\epsilon - \epsilon_\theta(\sqrt{\bar{\alpha}_t}x_0 + \sqrt{1-\bar{\alpha}_t}\epsilon, t)\right|\right|^2\right] \tag{4}$$

Additionally, we optimize the denoised output $\mu_\theta(x_t, t)$ with two extra terms. One loss term is based on the SSIM index, and the other is an edge-aware loss function, $\mathcal{L}_{\text{sobel}}$, derived from the Sobel operator [18] and extended for 3D inputs. Given the predicted denoised output $\hat{x}_0$ and the sample x_0 we convolve with the Sobel kernels (S) along each spatial axis to calculate gradient outputs.

$$G_x = S_x * x,\ G_y = S_y * x,\ G_z = S_z * x \tag{5}$$

Here, $*$ denotes 3D convolution, and $S_x, S_y, S_z \in \mathbb{R}^{3\times3\times3}$ are discrete 3D Sobel kernels that approximate the partial derivatives of the image along the x, y, and z axes. The gradient magnitudes are then approximated and combined:

$$\nabla x \approx \tanh(\sqrt{G_x^2 + G_y^2 + G_z^2}) \tag{6}$$

We use the tanh non-linearity to bound the gradients. We calculate the Sobel loss as:

$$\mathcal{L}_{\text{sobel}} = ||\nabla\hat{x}_0 - \nabla x_0|| \tag{7}$$

2.3 Bridging Network

At a predefined timestep K in the diffusion process, we perform a cross-domain transition from the noisy low-field sample x_K^{LF} to its high-field counterpart x_K^{HF}. While both are noised representations of paired scans, we observe a substantial mismatch in their latent distributions at this stage similar to Zhao et al. [19]. This discrepancy poses a challenge for the denoising process, which is trained solely on the high-field domain and expects inputs to lie within its data manifold.

To address this issue, we propose a dedicated bridge network that learns to translate x_K^{LF} into an intermediate representation that better aligns with the latent space of x_K^{HF}. This network is trained exclusively on noised paired samples and optimized to minimize the distance between its output and the corresponding high-field sample at timestep K. By explicitly modeling the domain gap at the noisy level, the bridge facilitates a smoother transition and enables the diffusion model to operate effectively in generating enhanced high-field-like outputs from low-field inputs.

We define the bridge network with parameters ϕ_{bridge}, which takes the noisy low-field input at timestep K, x_K^{LF}, and produces an improved latent representation aligned with the high-field domain:

$$\hat{x}_K^{HF} = x_K^{LF} + \phi_{\text{bridge}}(x_K^{LF}) \tag{8}$$

To train this network, we use the following loss terms. First, we enforce latent alignment by minimizing the ℓ_2 distance between the predicted representation $\hat{x}_K^{HF}$ and the true high-field sample x_K^{HF} at the same timestep. Second, we leverage the fact that x_K^{HF} is constructed by adding Gaussian noise ϵ_K to the clean high-field image x_0^{HF}. By subtracting the same noise from the predicted representation $\hat{x}_K^{HF}$, we recover an estimate of the clean high-field image, $\hat{x}_0^{HF}$. We then introduce an additional ℓ_2 loss term between $\hat{x}_0^{HF}$ and the ground truth x_0^{HF} to guide the bridge network toward producing representations more suitable to denoising. Finally, we regularize the bridge network with a third loss term that enforces a form of self-identity. After each standard training step, we input the high-field sample into the bridge and encourage the network to produce a minimal residual. Specifically, we use the high-field label as input and minimize the following identity loss:

$$\mathcal{L}_{\text{identity}} = ||x_K^{HF} - \left(x_K^{HF} + \phi_{\text{bridge}}(x_K^{HF})\right)||_2^2 \tag{9}$$

This term ensures that the bridge network does not unnecessarily change inputs already aligned with the high-field distribution, promoting stability and preventing overcorrection.

2.4 Inference

In addition to the conventional denoising diffusion inference algorithm, we find adding intermediate steps that further "correct" at each timestep, inspired from the Predictor-Corrector (PC) sampling of Chung et al. [20], increases denoising efficiency. We use the following algorithm. At each timestep, instead of denoising and adding the noise for the next timestep, we iterate on the same noise level with the same noise sample.

We define two operators to simulate the forward and reverse diffusion processes to simplify the pseudo-code given later in this subsection. **AddNoise** adds noise to a clean sample x_0 at timestep t. **RemoveNoise** denoises a noisy sample x_t at timestep t using the predicted noise $\epsilon_\theta(x_t, t)$:

$$\text{AddNoise}(x, \epsilon, t) = \sqrt{\bar{\alpha}_t} \cdot x + \sqrt{1 - \bar{\alpha}_t} \cdot \epsilon \tag{10}$$

$$\text{RemoveNoise}(x_t, \epsilon, t) = \frac{1}{\sqrt{\alpha_t}} \left(x_t - \frac{\beta_t}{\sqrt{1 - \bar{\alpha}_t}} \cdot \epsilon \right) \tag{11}$$

With all the components in place, we construct our full inference algorithm, as shown in Algorithm 1, for enhancing low-field samples to their high-field counterparts. We use $J = 3$ for the refinement loop of the algorithm.

3 Experiments and Results

We evaluated our method on the hidden validation set of the challenge by submitting enhanced pseudo high-field images generated from the corresponding low-field inputs. All performance metrics, unless otherwise stated, were computed by the challenge organizers. For all models, intensities of the input volumes were normalized to the range $[-1, 1]$.

Algorithm 1: Inference algorithm used. In succession the bridging and then the denoising diffusion produces the pseudo high-field sample $\hat{x}_0^{HF}$.

$\epsilon_K \sim \mathcal{N}(0, 1)$;
$x_K^{LF} \leftarrow \text{AddNoise}\left(x_0^{LF}, \epsilon_K, K\right)$;
$\hat{x}_K^{HF} \leftarrow x_K^{LF} + \phi_{\text{bridge}}(x_K^{LF})$;
$\hat{x}_0^{HF} \leftarrow \text{RemoveNoise}\left(\hat{x}_K^{HF}, \epsilon_K, K\right)$;
for $i = K, ..1,$ **do**
  $\epsilon_i \sim \mathcal{N}(0, 1)$;
  for $j = J, .., 1$ **do**
    $\hat{x}_i^{HF} \leftarrow \text{AddNoise}\left(\hat{x}_0^{HF}, \epsilon_i, i\right)$;
    $\hat{x}_0^{HF} \leftarrow \text{RemoveNoise}\left(\hat{x}_i^{HF}, \epsilon_\theta(\hat{x}_i^{HF}, i), i\right)$;
return $\hat{x}_0^{HF}$

3.1 Model Training

We trained our denoising diffusion model solely on high-field images. We combine all three contrasts along the channel dimension, enabling cross-contrast learning. The network architecture follows the U-Net design proposed by Karras et al. with [21]. We extended this network to work with 3D convolutions. This network was trained using the AdamW [22] optimizer with a learning rate of 0.0003. We used 3D patches of size $(128, 128, 32)$ for training. Since the high-field images from the hidden validation set were not available, we report performance on an internal test set of 5 scans. Using our proposed inference strategy, the model achieved an SSIM of 0.91 and a PSNR of 30.12 dB. These values represent the upper bound of achievable quality for our full pipeline since, this level simulates an ideal bridge as we stay in the high-field domain.

We trained the bridge network using the Swin UNETR [23] architecture, focusing on noise levels between timesteps 250 and 750 (out of 1000). We choose this wider spectrum both as a method for better generalization and to give us the room to alter the bridging point later on in the pipeline. For the final pipeline, we selected $K = 400$ as the bridging timestep. At this level, the bridge model achieved an average MSE of 0.05 between predicted and ground truth noisy high-field samples in out internal test set, indicating better latent space alignment under noise, since without the bridge model this average is 0.32.

3.2 Low-Field Enhancement

We combined the trained bridge network and high-field diffusion model to perform full enhancement of low-field MR images. The model produced pseudo high-field outputs from the low-field validation set samples, which were evaluated by the challenge organizers. Our model operated on 3D patches of size $(128, 128, 32)$, using a sliding window inference with an overlap of 0.7. Our method achieved an SSIM of 0.779 and a PSNR of 22.85 dB on the hidden validation set. This confirmed that the bridged diffusion approach significantly narrows the gap between low and high-field quality. An example of this is shown in Fig. 3.

3.3 Ablation Study

We conducted an ablation study to evaluate the contribution of each component:

1. Neither diffusion nor bridge model: Using low-field samples as high-field samples. Assuming $x_0^{LF} \approx x_0^{HF}$.
2. No diffusion model: Applying the bridge, removing the added noise perfectly. Given ϵ_K added to x_0^{LF} to make x_K^{LF}, we remove ϵ_K from $\phi_{\text{bridge}}\left(x_K^{LF}\right)$ without any diffusion.
3. No bridge model: Assuming $x_K^{LF} \approx x_K^{HF}$. Using the diffusion network to denoise x_K^{LF}.
4. Full pipeline: As described in Algorithm 1.

The results as calculated by the organizers are listed in Table 1.

Table 1. Ablation study showing the contribution of various components. All metrics are calculated by the challenge organizers on the secret validation set.

Model	Component		Metrics			
	Bridge	Diffusion	SSIM	PSNR	MAE	NMSE
1			0.569	17.544	0.072	0.234
2	✓		0.756	22.712	0.036	0.072
3		✓	0.601	17.602	0.07	0.231
4	✓	✓	**0.779**	**22.85**	**0.034**	**0.070**

4 Discussion and Conclusion

We presented a bridged denoising diffusion model that enhances low-field MR images by explicitly aligning their noisy representations with those of high-field images. By introducing a bridge network at a specific diffusion timestep, we enable a denoising diffusion model to effectively process low-field inputs into pseudo high-field outputs. Our ablation study shows that both the bridge and the diffusion components contribute: the bridge improves latent alignment, while the diffusion model provides high-quality reconstruction. Their combination leads to the highest performance. However, we observe that the contribution of the denoising diffusion model is relatively small compared to the bridge model. Further improvements in this regard should be made.

While our method relies on paired training data, future work could explore semi-supervised or unpaired approaches for broader applicability and data availability. Overall, our framework offers an effective solution for low-field MRI enhancement.

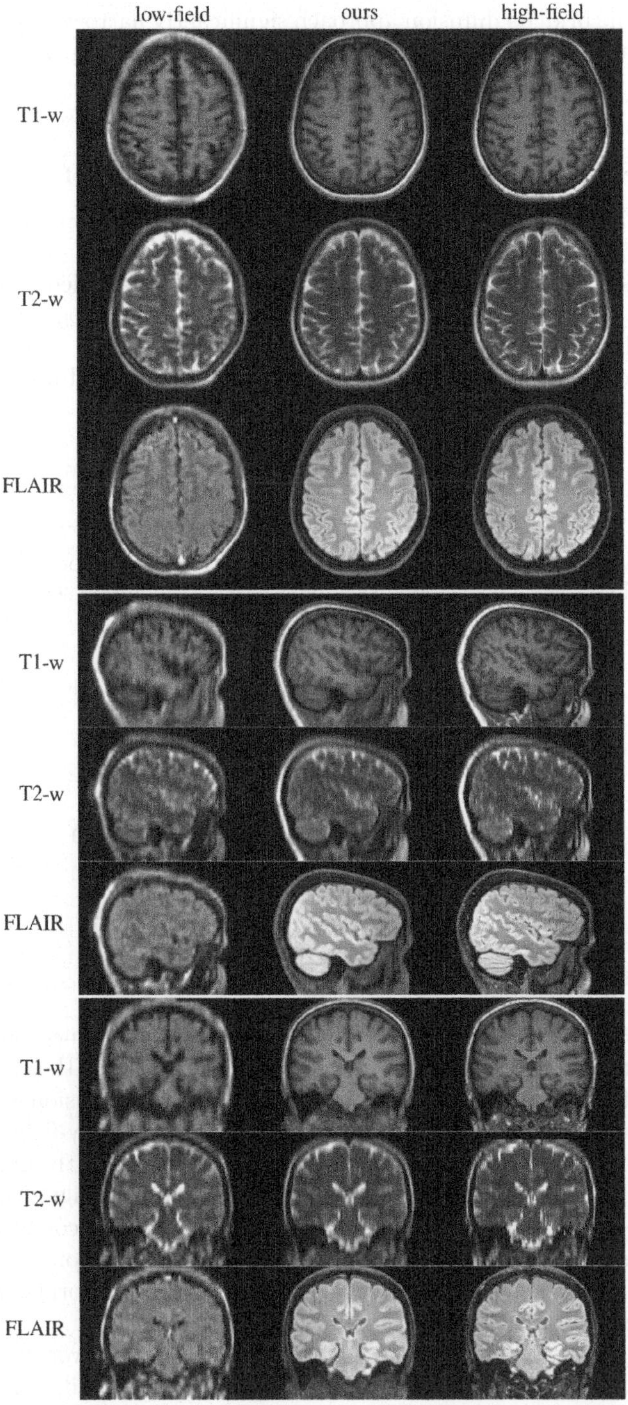

Fig. 3. Example result of the low-field enhancement. From left to right, low-field, pseudo high-field, and the ground truth high-field sample. Three views of the same scan are presented in all three contrasts.

Acknowledgments. This publication is part of the project AI4AI with file number P22.017 of the research program Perspectief which is (partly) financed by the Dutch Research Council (NWO) under the grant TTW Perspectief 2022–2023.

References

1. Dayarathna, S., et al.: Deep learning based synthesis of MRI, CT and PET: review and analysis. Med. Image Anal. **92**, 103046 (2024). http://dx.doi.org/10.1016/j.media.2023.103046. ISSN 1361-8415
2. Sarracanie, M., Salameh, N.: Low-field MRI: how low can we go? A fresh view on an old debate. Front. Phys. **8** (2020). http://dx.doi.org/10.3389/fphy.2020.00172. ISSN 2296-424X
3. Arnold, T.C., Freeman, C.W., Litt, B., Stein, J.M.: Low-field MRI: clinical promise and challenges. J. Magn. Reson. Imaging **57**, 25–44 (2023). https://onlinelibrary.wiley.com/doi/10.1002/jmri.28408. ISSN 1053-1807, 1522-2586
4. Webb, A., O'Reilly, T.: Tackling SNR at low-field: a review of hardware approaches for point-of-care systems. Magn. Reson. Mater. Phys. Biol. Med. **36**, 375–393 (2023). https://link.springer.com/10.1007/s10334-023-01100-3. ISSN 1352-8661
5. Guallart-Naval, T., et al.: Portable magnetic resonance imaging of patients indoors, outdoors and at home. Sci. Rep. **12**, 13147 (2022). https://www.nature.com/articles/s41598-022-17472-w. ISSN 2045-2322
6. Deoni, S.C., et al.: Neuroimaging and verbal memory assessment in healthy aging adults using a portable low-field MRI scanner and a web-based platform: results from a proof-of-concept population-based cross-section study. Brain Struct. Funct. **228**, 493–509 (2023). ISSN 18632661
7. Yuen, M.M., et al.: Portable, low-field magnetic resonance imaging enables highly accessible and dynamic bedside evaluation of ischemic stroke. Sci. Adv. **8**, 3952 (2022). https://www.science.org/doi/10.1126/sciadv.abm3952. ISSN 2375-2548
8. Okar, S.V., et al.: Portable ultra-low-field MRI for progressive multifocal leukoencephalopathy: case studies, sensitivity, and potential applications. J. Neurol. **272**, 193 (2025). https://link.springer.com/10.1007/s00415-025-12938-z. ISSN 0340-5354
9. Rowand, E., Owusu, R., Sibole, A., Abu-Haydar, E., Delarosa, J.: Feasibility and usability of low-field magnetic resonance imaging for pediatric neuroimaging in low- and middle-income countries: a qualitative study. Med. Dev. Evid. Res. **18**, 107–121 (2025). https://www.dovepress.com/feasibility-and-usability-of-low-field-magnetic-resonance-imaging-for--peer-reviewed-fulltext-article-MDER. ISSN 1179-1470
10. Islam, K.T., et al.: Improving portable low-field MRI image quality through image-to-image translation using paired low- and high-field images. Sci. Rep. **13** (2023). http://dx.doi.org/10.1038/s41598-023-48438-1. ISSN 2045-2322
11. Lin, H., et al.: Low-field magnetic resonance image enhancement via stochastic image quality transfer. Med. Image Anal. **87**, 102807 (2023). http://dx.doi.org/10.1016/j.media.2023.102807. ISSN 1361-8415
12. Chung, H., Ye, J.C.: Score-based diffusion models for accelerated MRI. Med. Image Anal. **80**, 102479 (2022). http://dx.doi.org/10.1016/j.media.2022.102479. ISSN 1361-8415

13. Dayarathna, S., et al.: McCaD: multi-contrast MRI conditioned, adaptive adversarial diffusion model for high-fidelity MRI synthesis. In: 2025 IEEE/CVF Winter Conference on Applications of Computer Vision (WACV), pp. 670–679. IEEE (2025). http://dx.doi.org/10.1109/WACV61041.2025.00075
14. Lin, X., et al.: Medical Image Computing and Computer Assisted Intervention – MICCAI 2024 775–785. Springer, Cham (2024). http://dx.doi.org/10.1007/978-3-031-72104-5_74. ISBN 9783031721045
15. Dayarathna, S., Islam, K.T., Chen, Z.: Ultra low-field to high-field MRI translation using adversarial diffusion. In: 2024 IEEE International Symposium on Biomedical Imaging (ISBI), pp. 1–4 (2024)
16. Wang, Z., Bovik, A., Sheikh, H., Simoncelli, E.: Image quality assessment: from error visibility to structural similarity. IEEE Trans. Image Process. **13**, 600–612 (2004)
17. Ho, J., Jain, A., Abbeel, P.: Denoising diffusion probabilistic models (2020). https://arxiv.org/abs/2006.11239
18. Kanopoulos, N., Vasanthavada, N., Baker, R.: Design of an image edge detection filter using the Sobel operator. IEEE J. Solid-State Circuits **23**, 358–367 (1988)
19. Zhao, K., Hung, A.L.Y., Pang, K., Zheng, H., Sung, K.: PartDiff: image super-resolution with partial diffusion models (2023). https://arxiv.org/abs/2307.11926
20. Chung, H., Ye, J.C.: Score-based diffusion models for accelerated MRI (2021). https://arxiv.org/abs/2110.05243
21. Karras, T., et al.: Analyzing and improving the training dynamics of diffusion models (2023). https://arxiv.org/abs/2312.02696
22. Loshchilov, I., Hutter, F.: Decoupled weight decay regularization (2017). https://arxiv.org/abs/1711.05101
23. Hatamizadeh, A., et al.: Swin UNETR: swin transformers for semantic segmentation of brain tumors in MRI images (2022). https://arxiv.org/abs/2201.01266

LowDM: A Diffusion-Based Deep Learning Framework for Generating High-Field Quality Images from Portable Low-Field MRI

Alfredo Lucas[1(✉)], Chetan Vadali[1], and Joel M. Stein[1,2]

[1] Perelman School of Medicine, University of Pennsylvania, Philadelphia, USA
alfredo.lucas@pennmedicine.upenn.edu

[2] Department of Radiology, University of Pennsylvania, Philadelphia, USA

Abstract. Portable ultra-low-field strength (64 mT) MRI scanners make bedside neuroimaging feasible, but lower image quality compared to conventional high-field systems may restrict diagnostic use. Here we introduce LowDM, a multi-contrast conditional Wavelet Diffusion Model that converts 64 mT volumes into 3 T-like images. Fifty paired brain MRI datasets from the 2025 Ultra-Low-Field MRI Image Enhancement Challenge were used for training; 10 additional cases served for validation. During training, discrete wavelet transforms of 3 T targets were progressively noised over 1000 steps, and a 3D UNet learned to reverse this process while being conditioned on co-registered 64 mT coefficients. Inference required only 50 denoising steps to deliver synthetic 3 T images. Compared to native 64 mT scans, LowDM outputs achieved higher structural fidelity (SSIM 0.705 vs 0.448) and signal quality (PSNR 28.8 dB vs 20.0 dB) while reducing MAE and NMSE by more than half. Volumetric analyses with SynthSeg confirmed superior preservation of cortical and subcortical anatomy. These results demonstrate that diffusion-based enhancement can help bridge the quality gap between portable and conventional high-field MRI, potentially expanding access to clinically useful neuroimaging in underserved environments.

Keywords: Low-field MRI · Deep Learning · Diffusion Model

1 Introduction

Conventional high-field magnetic resonance imaging (MRI) scanners (1.5 T or higher) are expensive, immobile, and resource intensive, which limits their accessibility, particularly in low-resource settings or remote locations. Portable low-field magnetic resonance scanners operating at ultra-low field strengths (e.g. 64 mT) can improve access due to their lower cost, increased portability, and simplified operational requirements [1,2,11]. Despite these advantages, lower tissue

A. Lucas and C. Vadali—These authors contributed equally to this work.

Z. Chen et al. (Eds.): ULF-EnC 2025, LNCS 16293, pp. 67–75, 2026.
https://doi.org/10.1007/978-3-032-23344-8_8

contrast and spatial resolution in images produced by portable low-field scanners may hinder their use for clinical diagnosis and research applications.

Deep learning architectures such as generative adversarial networks (GANs) have shown promise in addressing image quality limitations in the post-acquisition image space by transforming low-field images into high-field-like images [9,10]. LowGAN [10], a multi-contrast GAN-based framework to generate 3 T images from 64 mT images, was recently introduced. LowGAN improved overall image quality, as well as volumetric measurements derived from the low-field images, making them comparable to those derived from high-field inputs. Nevertheless, LowGAN still faced some limitations, including the use of an older 2-dimensional architecture [13].

Diffusion models have recently outperformed GAN-based approaches in various medical image enhancement tasks, offering improved stability, fidelity, and structural preservation [8,12]. In these models, a deep learning architecture (e.g., UNet) is trained to predict the noise in an input in order to denoise the input through a reverse diffusion process. In the absence of input conditioning volumes, the reverse denoising process generates images without constraints. If input conditioning volumes are included during training and inference (i.e., conditional diffusion), the denoising process can be constrained to generate images based on those conditioning volumes (e.g. high-field images from low-field conditioning volumes).

To overcome the limitations of our prior work, and build on the success of diffusion models, we developed LowDM, a multi-contrast diffusion-based deep learning framework for high-field image synthesis from low-field inputs. LowDM employs a conditional Wavelet Diffusion Model (cWDM) where the diffusion process is trained in the wavelet domain, using low-field images as conditioning images and high-field images as target outputs [8]. This approach has been previously successfully applied for missing-contrast synthesis [8], and in this work, we expand its application to low-to-high-field image synthesis.

2 Methods

2.1 Dataset

The data for this study was obtained from the *2025 - Ultra-Low-Field MRI Image Enhancement Challenge (ULF-EnC)*. ULF-EnC utilizes a dataset comprising paired ultra-low-field (64 mT) and high-field (3 T) MRI scans collected using a Hyperfine Swoop (64 mT) scanner and a Siemens Biograph mMR (3 T) scanner at Monash Biomedical Imaging, Monash University, Australia. Standardized procedures were followed by trained radiologists to acquire T1-weighted (T1w), T2-weighted (T2w), and fluid attenuated inversion recovery (FLAIR) brain MRI scans in healthy volunteers. Detailed imaging protocols, including resolution, matrix size, repetition time (TR), echo time (TE), and inversion time (TI), are provided in the original source documentation [5–7,9].

The dataset consists of 50 training cases, 10 validation cases, and 15 test cases. The training set was intentionally larger to facilitate robust model learning, while the validation set was used for hyperparameter tuning. The test set was withheld from participants; therefore, results from that set are not included in this manuscript. The competition-provided validation set did not include 3 T images, and metrics from that validation set were directly computed by the program organizers. Results from our analyses are presented from an internal validation set (8 held-out subjects from the 50 training subjects provided), the competition validation set (10 subjects), and the competition test set (15 subjects).

2.2 Preprocessing

Privacy was maintained by applying identical face removal masks to both 64 mT and 3 T paired scans. Additional data preprocessing performed by the organizers of the challenge included manual quality control checks, conversion from DICOM to NIFTI format, standardization to a 1 mm^3 isotropic resolution, bias field correction using FSL-FAST, and co-registration of scans using FLIRT to enable accurate comparisons. The 3 T scans served as the reference standard for evaluating the quality of enhanced 64 mT images. The only additional preprocessing done here prior to training and inference was quantile clipping of the images between $1st$ and $99th$ percentile followed by Min-Max scaling the images between 0 and 1.

2.3 Data Augmentation

Data augmentation was applied during the training phase to improve model robustness and generalizability. When each subject was loaded for training, random affine and elastic transformations were applied to the input images to simulate variability in input data. Random affine transformations included translation (probability = 0.5, maximum ±15 pixels), shearing (probability = 0.5, maximum ±15°), and rotation (probability = 0.5, maximum ±25°). Additionally, elastic deformation was applied with a probability of 0.2, characterized by magnitude (α) values ranging from 0 to 2 and smoothness (σ) values between 6 and 10. Augmentation parameters were randomly determined once per subject to ensure consistency across all image slices within that subject, maintaining anatomical coherence.

2.4 Diffusion Model Architecture

We trained a 3-dimensional (3D) conditional Wavelet Diffusion Model (cWDM) for low-to-high field image translation (Fig. 1), based on the cWDM from Friedrich et al. [8]. The model parameters included 128 base channels, 8 residual blocks, and 4 attention heads with $16 \times 16 \times 16$ resolution. During training, 3D 3 T images were reshaped into a 3-channel input to include T1w, T2w, and FLAIR contrasts. We then apply a discrete wavelet transform (DWT) using Haar wavelets to the multi-contrast 3 T input, decomposing each input into 8 wavelet sub-bands, and apply Gaussian noise iteratively in the wavelet domain over 1000

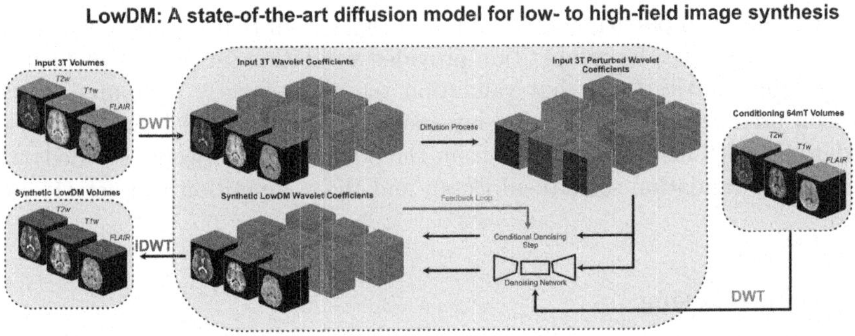

Fig. 1. Schematic of the LowDM architecture - The model performs diffusion-based denoising in the wavelet domain. During training, input 3 T images undergo discrete wavelet transform (DWT), perturbation, and iterative denoising conditioned on 64 mT reference volumes. The final synthetic output is reconstructed via inverse DWT (iDWT), resulting in a 3 T-like image across T1w, T2w, and FLAIR contrasts.

timesteps until the input has been made indistinguishable from noise. Using the paired 64 mT images, also reshaped into a 3-channel multi-contrast input and then converted to wavelet domain with DWT, we constrain the diffusion denoising process with these wavelet coefficients, and a UNet neural network learns the parameters required for the denoising diffusion process conditioned by the 64 mT inputs. We compute a mean squared error (MSE) loss for each ground truth (3 T) and output (synthetic) wavelet coefficient and iterate this process during training. During inference, only the 64 mT T1w, T2w, and FLAIR 3D conditioning volumes are provided, and using the learned parameters for denoising over 50 timesteps, LowDM returns synthetic 3 T-like T1w, T2w, and FLAIR volumes.

2.5 Training

The model was trained in the method described above with a batch size of 1, learning rate of 1×10^{-5}, and dropout of 0.1. For the first 60,000 iterations, the training set was divided into 42 training subjects and 8 held-out validation subjects. This internal validation set was then used to compute preliminary results, shown in Figs. 3 and 4, in the early stages of the competition. We subsequently trained a model with all 50 training subjects for 800,000 iterations, and tested on the competition-provided validation (Fig. 2 and Table 1) and testing sets (Table 1). The pretrained LowDM model will be made available on: https://github.com/penn-cnt/LowDM.

2.6 Postprocessing

After inference, the outputs were post-processed to improve visual image contrast and smoothness, and to maximize image quality metrics including structural similarity index measure (SSIM), peak signal-to-noise ratio (PSNR), and

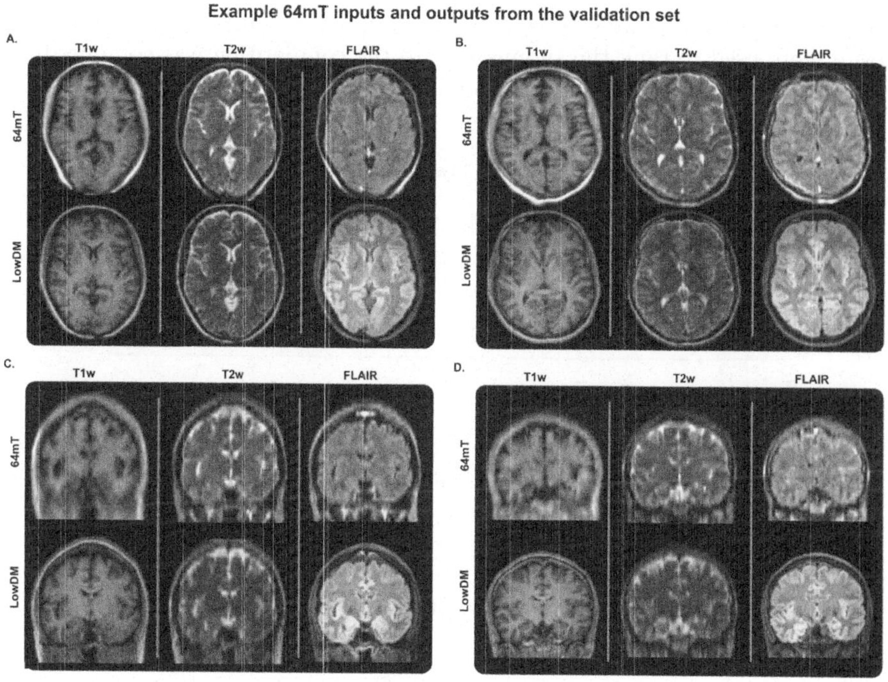

Fig. 2. Representative outputs from the validation set - T1w, T2w and FLAIR outputs from LowDM (bottom) compared to their corresponding 64 mT inputs (top) for two subjects in the competition provided validation set (POCEMR058 and POCEMR072). Note that there was no ground truth 3 T label for these subjects provided by the organizers. Panels A. and B. show representative axial slices across contrasts. Panels C. and D. show representative coronal slices across contrasts.

MSE in the reconstructed image domain. The post-processing pipeline consisted of several sequential steps. First, total variation (TV) denoising was applied as an edge-preserving smoother with a weighting factor of 0.01. Then, mild image sharpening was performed using an unsharp mask with a radius of 3.0 pixels and a sharpening amount of 0.1. Following TV denoising, contrast enhancement was performed using 3-dimensional Contrast Limited Adaptive Histogram Equalization (CLAHE) with a clip limit of 0.001. Next, robust range clipping was executed by restricting pixel intensities to the 0.5th and 99.5th percentiles, followed by normalization back to the [0, 1] intensity range. Finally, a Gaussian filter ($\sigma = 0.8$) was applied to mitigate small speckles and further refine image smoothness. Using these postprocessed images, SSIM, PSNR, mean absolute error (MAE), and normalized mean squared error (NMSE) were computed both in the internal validation set (Fig. 3), as well as in the competition-provided validation and test sets (Table 1). Of note, the image quality metrics computed by the competition organizers were done on whole-brain masked images. The final composite score used to rank the submissions to the competition was computed as: $CompositeScore = 0.7\,\text{SSIM} + 0.1\,(\text{PSNR}/32) + 0.1\,(1 - \text{MAE}) + 0.1\,(1 - \text{NMSE})$.

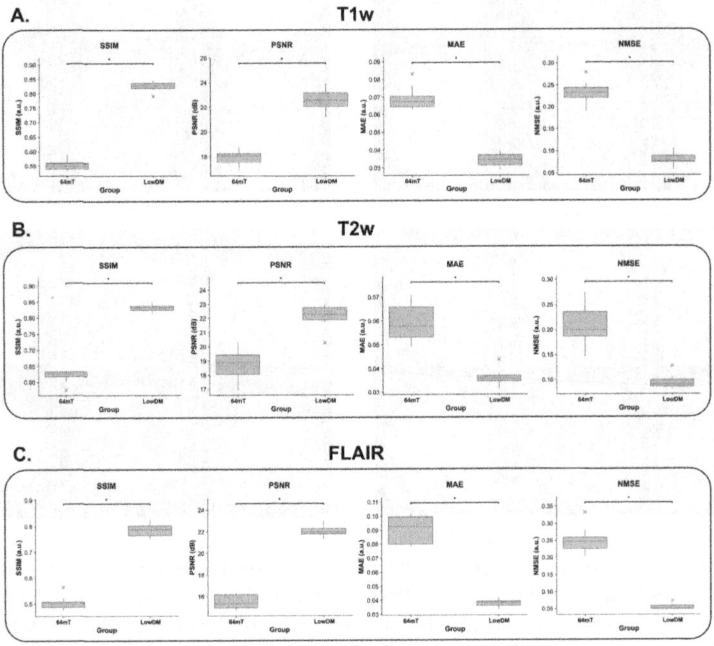

Fig. 3. Image Quality - Quantitative image quality metrics in the internal validation set. SSIM, PSNR, MAE, and NMSE were computed between 3 T and 64 mT, and between 3 T and LowDM outputs for T1w (Panel A), T2w (Panel B), and FLAIR (Panel C) contrasts. * denotes $p < 0.05$ after Bonferroni correction

3 Results

3.1 LowDM Generates Outputs that Are Visually Similar to 3 T Images

From the 64 mT inputs, LowDM generated T1w, T2w, and FLAIR output images that closely resemble 3 T images. Visual inspection (Fig. 2) demonstrated superior image quality with LowDM compared to unprocessed 64 mT images, particularly in the cortical ribbon and subcortical structures. The synthetic outputs also remove flow-related venous hyperintense signals present on 64 mT T2w and FLAIR inputs but not 3 T target images.

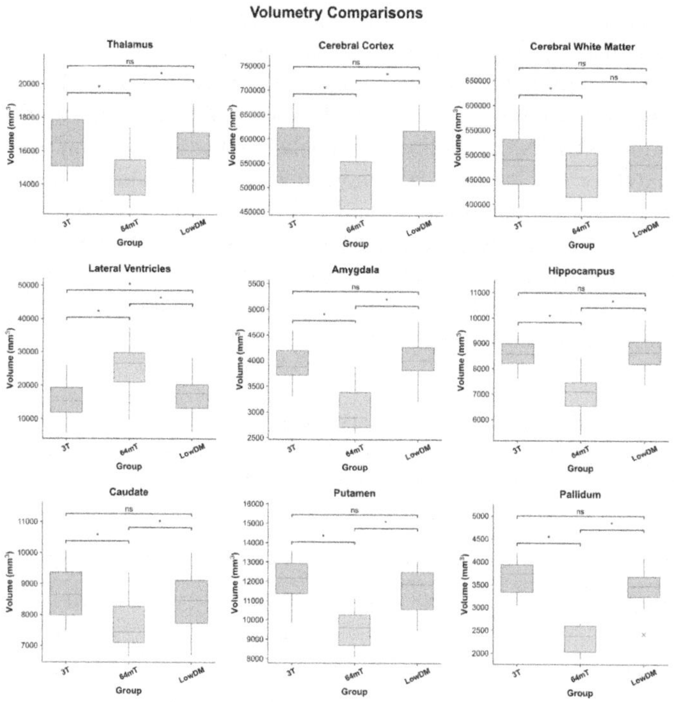

Fig. 4. Cortical and Subcortical Volumetry - Regional volumes were compared between 3 T, 64 mT, and LowDM images for subjects in the internal validation set. * denotes $p < 0.05$ after Bonferroni correction

3.2 Image Quality Metrics

LowDM synthetic 3 T images quantitatively resemble true 3 T images. LowDM-generated images demonstrated improved quality in the internal validation set compared to the input 64 mT images (Fig. 3), with greater SSIM and PSNR and lesser MAE and NMSE for T1w, T2w, and FLAIR contrasts compared to 64 mT images.

3.3 Volumetry

Synthetic LowDM images generated from the internal validation set also outperformed 64 mT inputs in capturing regional brain volumes, measured with SynthSeg [3,4], when compared to reference 3 T images (Fig. 4).

4 Discussion

LowDM addresses some of the inherent limitations from lower signal to noise in low-field imaging. Using a 3D conditional Wavelet Diffusion Model, LowDM generates multi-contrast synthetic high-field MRI images from ultra-low-field inputs, significantly improving visual clarity and anatomical fidelity across multiple contrasts (T1-weighted, T2-weighted, and FLAIR). LowDM demonstrated robustness to unseen data, as exemplified by comparable performance between validation and test sets, as well as 3-dimensional volumes that did not have significant patch or tile artifacts common in 2- or 2.5-dimensional models.

By using both a diffusion model and a 3-dimensional architecture, LowDM provides a state-of-the-art approach for low-to-high-field image synthesis. Nevertheless, our results highlight the importance of appropriate postprocessing during image synthesis, as it can provide significant perceptual and quantitative quality gains at little computational expense.

Table 1. Image quality metrics for LowDM outputs in the competition-provided validation set before (BP) and after (AP) postprocessing, and in the final test set after postprocessing. These metrics were computed directly by the organizers of the competition.

Metric	Validation (BP)	Validation (AP)	Test (AP)
SSIM	0.448	0.705	0.710
PSNR	19.962	28.768	29.629
MAE	0.186	0.072	0.071
NMSE	0.739	0.160	0.066
Composite Score	**0.483**	**0.760**	**0.776**

Despite encouraging results, this work utilized a well-curated dataset of healthy participants. Validation on larger and more diverse clinical cohorts, including cases with pathologic lesions [10], will be essential to fully understand the clinical applicability and limitations of LowDM in real-world diagnostic settings.

In conclusion, LowDM offers a state-of-the-art approach to improve the diagnostic utility of portable ultra-low-field MRI scanners, potentially expanding MRI accessibility in underserved or remote settings.

Disclosure of Interests. JMS: Payments to the institution from Hyperfine Research through a research agreement; consulting fees from Subtle Medical (related to deep learning image quality improvement) and Centaur Diagnostics (related to medical image annotation).

References

1. Arnold, T.C., Freeman, C.W., Litt, B., Stein, J.M.: Low-field MRI: clinical promise and challenges. J. Magn. Reson. Imaging **57**(1), 25–44 (2023). https://doi.org/10.1002/jmri.28408

2. Arnold, T.C., et al.: Sensitivity of portable low-field magnetic resonance imaging for multiple sclerosis lesions. NeuroImage Clin. **35**, 103101 (2022). https://doi.org/10.1016/j.nicl.2022.103101
3. Billot, B., et al.: SynthSeg: segmentation of brain MRI scans of any contrast and resolution without retraining. Med. Image Anal. **86**, 102789 (2023). https://doi.org/10.1016/j.media.2023.102789
4. Billot, B., Magdamo, C., Cheng, Y., Arnold, S.E., Das, S., Iglesias, J.E.: Robust machine learning segmentation for large-scale analysis of heterogeneous clinical brain MRI datasets. Proc. Nat. Acad. Sci. **120**(9), e2216399120 (2023). https://doi.org/10.1073/pnas.2216399120
5. Dayarathna, S., Islam, K.T., Chen, Z.: Ultra low-field to high-field MRI translation using adversarial diffusion. In: 2024 IEEE International Symposium on Biomedical Imaging (ISBI), pp. 1–4. IEEE (2024). https://doi.org/10.1109/ISBI56570.2024.10635808
6. Dayarathna, S., Islam, K.T., Uribe, S., Yang, G., Hayat, M., Chen, Z.: Deep learning based synthesis of MRI, CT and pet: review and analysis. Med. Image Anal. **92**, 103046 (2024). https://doi.org/10.1016/j.media.2023.103046
7. Dayarathna, S., et al.: McCaD: multi-contrast MRI conditioned, adaptive adversarial diffusion model for high-fidelity MRI synthesis. In: 2025 IEEE/CVF Winter Conference on Applications of Computer Vision (WACV), pp. 670–679. IEEE (2025). https://doi.org/10.1109/WACV61041.2025.00075
8. Friedrich, P., Durrer, A., Wolleb, J., Cattin, P.C.: cWDM: conditional wavelet diffusion models for cross-modality 3d medical image synthesis. arXiv preprint arXiv:2411.17203 (2024). https://doi.org/10.48550/arXiv.2411.17203
9. Islam, K.T., et al.: Improving portable low-field MRI image quality through image-to-image translation using paired low- and high-field images. Sci. Rep. **13**(1), 21183 (2023). https://doi.org/10.1038/s41598-023-48438-1
10. Lucas, A., et al.: Multisequence 3-T image synthesis from 64-mt low-field-strength MRI using generative adversarial networks in multiple sclerosis. Radiology **315**(1), e233529 (2025). https://doi.org/10.1148/radiol.233529
11. Mazurek, M.H., et al.: Portable, bedside, low-field magnetic resonance imaging for evaluation of intracerebral hemorrhage. Nat. Commun. **12**(1), 1 (2021). https://doi.org/10.1038/s41467-021-25441-6
12. Rombach, R., Blattmann, A., Lorenz, D., Esser, P., Ommer, B.: High-resolution image synthesis with latent diffusion models. arXiv preprint arXiv:2112.10752 (2021)
13. Wang, S., Zaharchuk, G.: Generative artificial intelligence to improve ultralow-field-strength MRI scan quality. Radiology **315**(1), e250932 (2025). https://doi.org/10.1148/radiol.250932

Super-Resolution of Ultra-Low-Field MRI Using a GAN-Based Visual Transformer Network

Aram Salehi[1,2], Tavia E. Evans[2,3], Chloé Najac[1], Beatrice Lena[1], Yiming Dong[1], Ruben van den Broek[1], Hieab H. H. Adams[2,4], and Andrew Webb[1](✉)

[1] C.J. Gorter MRI Center, Department of Radiology, Leiden University Medical Center, Leiden, The Netherlands
{a.salehi,c.f.najac,b.lena,y.dong1,r.b.van_den_broek, A.Webb}@lumc.nl

[2] Department of Human Genetics, Radboud University Medical Center, Nijmegen, The Netherlands
{Tavia.Evans,Hieab.Adams}@radboudumc.nl

[3] Global Brain Health Institute, Trinity College Dublin, Dublin, Ireland

[4] Latin American Brain Health Institute, Universidad Adolfo Ibañez, Santiago de Chile, Chile

Abstract. Ultra low-field (ULF) magnetic resonance imaging (MRI) is a cost-effective and portable imaging modality; however, it has lower spatial resolution and reduced signal-to-noise ratio compared to high-field MRI due to the lower magnetic field strength. As a result, some anatomical details may be lost due to lower contrast-to-noise and higher partial volume effects. To help address this issue, we propose a generative adversarial network (GAN)-based deep learning approach to enhance the resolution of the ULF-MR images. Our model, built on visual transformers, is trained on randomly extracted patches from T1-weighted, T2-weighted, and fluid attenuated inversion recovery (FLAIR) ULF-MR images to improve resolution across all three sequences. Using a paired low-/high-field MRI dataset provided by the Ultra-Low-Field MRI Image Enhancement Challenge (ULF-EnC), we trained and tested the model. The deep learning reconstructed images achieved a structural similarity (SSIM) score of 0.886 and a peak signal-to-noise ratio (PSNR) of 25.47.

Keywords: Ultra low-field MRI · generative adversarial network · super-resolution · image enhancement

1 Introduction

Magnetic resonance imaging (MRI) provides high-contrast images of soft tissues. However, high costs related to purchasing, acquiring, operating, and maintaining these systems limit their availability to healthcare institutions in high-resource areas [1]. As a result, MRI technology is unavailable or restricted in many global regions. To improve accessibility, there has been renewed interest in ultra-low-field MRI (ULF-MRI), with

Z. Chen et al. (Eds.): ULF-EnC 2025, LNCS 16293, pp. 76–83, 2026.
https://doi.org/10.1007/978-3-032-23344-8_9

field strength < 0.1 T [2, 3]. This offers a significant cost reduction, improved portability, and lower infrastructure requirements, thereby enabling broader access, particularly in low-resource settings. However, this improvement comes at the expense of a lower signal-to-noise ratio (SNR), due to the reduced strength of the magnetic field used in ULF-MRI systems compared to standard clinical systems, making image quality enhancement techniques highly desirable to improve diagnostic confidence [4, 5].

Deep learning (DL) approaches have been used in many recent studies to improve image quality in different domains [6]. These studies have employed convolutional neural networks (CNNs) [7], visual transformers [8], generative adversarial networks (GANs) [9], and diffusion models [10]. For instance, Chen et al. developed DCSRN, a super-resolution model based on densely connected convolutional networks for 3D MR brain images [11]; Iglesias et al. developed SynthSR, a U-Net-based model for generating high-resolution isotropic images from clinical scans [12]; Laguna et al. applied a CycleGAN-inspired architecture to integrate denoising, super-resolution [13]; and Islam et al. developed LoHiResGAN, an image-to-image translation model trained on paired low-field and high-field (HF) MRI data to enhance image quality and generate synthetic HF images from portable low-field scans [14]. These models are designed to use only a single sequence as the input to enhance the resolution of that same sequence, rather than leveraging complementary information from multiple sequences, which form the complete imaging protocol.

Recently, new models based on visual transformers have been proposed. Notably, Hatamizadeh et al. introduced Swin UNETR, a transformer-based framework designed for effective volumetric medical image segmentation [15]. The primary use of this network is for medical image segmentation, and to the best of our knowledge, it has not yet been used for super-resolution tasks. Its transformer-based architecture is capable of detecting patterns across entire 3D images, which allows missing details to be filled in and overall image resolution to be improved.

In this work, we combined two main ideas. First, we implemented a GAN framework in which, unlike typical GANs that use a U-Net as the generator backbone, we adopted a visual transformer architecture (Swin UNETR) for our generator. Second, we made use of three MRI contrasts—T1-weighted, T2-weighted, and fluid attenuated inversion recovery (FLAIR). By combining these different types of images, the network can draw on a wider range of information, which helps to improve the overall quality of the output. With this setup, our goal was to show that bringing together a transformer-based architecture with data from multiple sequences can lead to better results for ULF MRI reconstruction than using only one modality or more conventional generator models.

2 Method

2.1 Network

The current model utilizes a GAN network structure, as displayed in Fig. 1. The main differences from previous work are that the typical U-Net model in GAN was replaced with the Swin UNETR model, a 3D PatchGAN-style discriminator was used, and an additional ResNet3D-based enhancer was incorporated to remove patch boundary artifacts.

The generator is built on the Swin UNETR backbone, a vision transformer architecture specifically adapted for 3D medical imaging tasks. The model's inputs are three-channel patches of size 128 × 128 × 128. Each channel corresponds to co-registered T1-weighted, T2-weighted, and FLAIR images concatenated together. The generator outputs are the same-size image patches with the same number of channels to generate synthetic high-resolution images. A sigmoid activation function was added to the last layer of the output of the network to normalize the output voxel intensities to the [0, 1] range. For validation during the training, the full 3D images are reconstructed using patches with 70% overlap. The reconstructed images are then used to calculate losses of the generator and discriminator to ensure that patch-level predictions are aligned with the global image structure. The training was performed with a batch size of 2 for a total of 5000 epochs.

The discriminator is a 3D PatchGAN (Fig. 1). It is a combination of sequential 3D convolutional, batch normalization, and LeakyReLU layers. Its role is to distinguish between the generator's output and the ground-truth HF images, focusing on local structural details. The final layer applies a sigmoid activation to produce a patch-level real/fake probability map.

For training, we used a combination of pixel-wise L1 loss, SSIM loss, and a gradient difference loss, encouraging both voxel-wise accuracy and structural fidelity. The generator loss:

$$L_{gen} = \alpha L_1 + (1 - \alpha/2)L_{SSIM} + (1 - \alpha/2)L_{TV} \quad (1)$$

where L_1 is the mean absolute error, L_{SSIM} the 3D structural similarity loss, α is the weight of each loss which is equal to 0.5, and L_{TV} the total variation loss that promotes smoothness by penalizing large intensity differences between neighboring voxels in the predicted image.

The adversarial part of the framework uses a binary cross-entropy loss to compare the discriminator's output against the expected real or fake labels as:

$$L_{adv} = -E_{\{x \sim p_{\{real\}}\}}\left[log\, D(x)\right] - E_{\left\{\widehat{(x)} \sim p_{\{gen\}}\right\}}\left[log(1 - D(\hat{x}))\right] \quad (2)$$

where $D(x)$ denotes the discriminator output probability a, x represents the real sample from the data distribution $p_{\{real\}}$, and $\hat{x}$ represents generated samples from the model distribution $p_{\{gen\}}$.

Both the generator and discriminator are trained using the AdamW optimizer. Throughout training, paired 3D image patches are used, with data augmentation strategies such as random cropping, elastic deformation, and intensity scaling applied to improve model robustness.

After training the generator within the GAN framework, we trained an Enhancer (ResNet3D block (as shown in Fig. 1)) on the reconstructed full 3D images to remove patch boundary artifacts and enhance the overall image quality. For this second-stage training, we used a combination of mean absolute error (MAE) and SSIM as the loss function, assigning equal weights to both components.

Training was performed on the Snellius supercomputer using a single Phase 1 GPU node equipped with four NVIDIA A100 GPUs (each with 40 GiB HBM memory); only one GPU was used for the experiments.

All model development and training steps were implemented in Python using the PyTorch [16] and MONAI [17] libraries.

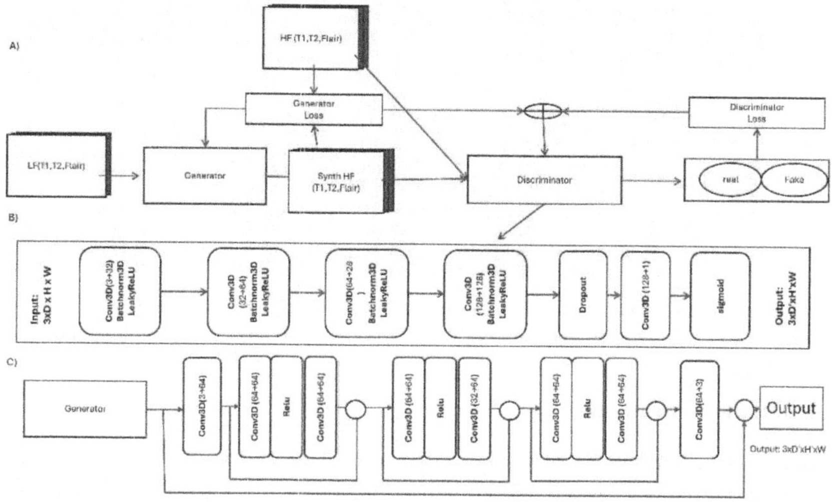

Fig. 1. Scheme of the proposed network architecture. (A) shows the hierarchical architecture based on Swin GAN. (B) shows the architecture of the discriminator. (C) shows the architecture of the Enhancer

2.2 Dataset

In this study, we used the Ultra-Low-Field MRI Image Enhancement Challenge (ULF-EnC) dataset, which consists of paired ULF (64 mT, Hyperfine Swoop) and HF (3 T, Siemens Biograph mMR) brain MRI scans acquired at Monash Biomedical Imaging. Each subject in the dataset includes T1-weighted, T2-weighted, and FLAIR images from both scanners. The dataset contains 50 subjects for training, of which 90% (45 subjects) are used for training and 10% (5 subjects) for validation in this work. Model performance was evaluated on 10 subjects from the test set which HF pairs are not provided, in line with the challenge's evaluation protocol. An example of input and ground truth of the dataset is shown in Fig. 2.

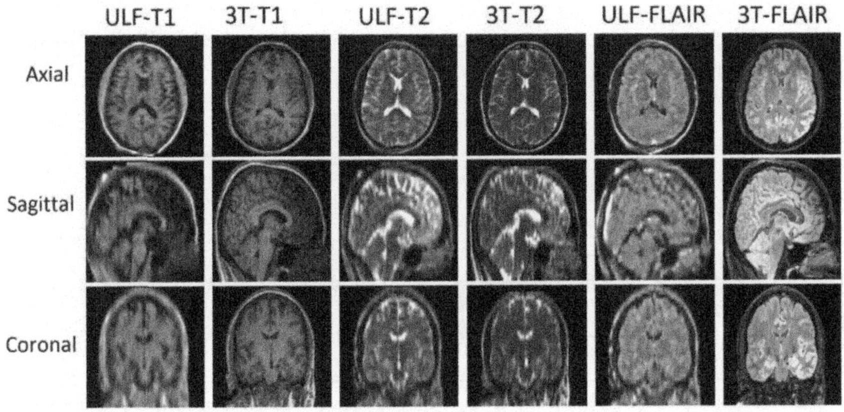

Fig. 2. Representative slices from T1-weighted, T2-weighted, and FLAIR MRI scans acquired at ULF (64 mT) and HF (3 T). For each modality and field strength, axial, sagittal, and coronal views are displayed.

2.3 Evaluation Parameters

To evaluate the performance of the proposed model, several widely used quantitative metrics were employed, namely the Structural Similarity Index Measure (SSIM), Peak Signal-to-Noise Ratio (PSNR), Mean Absolute Error (MAE), and Normalized Mean Squared Error (NMSE). Together, these metrics provide a comprehensive evaluation of image quality and model performance.

3 Results

Quantitative evaluation metrics for the input ULF MR and DL-reconstructed MR images from the proposed model, compared to the reference (HF MRI images, validation set), are summarized in Table 1. The mean SSIM for DL-reconstructed vs. reference was 0.886 ± 0.013, markedly higher than that for ULF vs. reference (0.543 ± 0.060). The PSNR also increased from 17.295 ± 1.619 for ULF to 25.466 ± 0.613 for the DL-reconstructed output, which reflects greater image fidelity with respect to the reference. In addition, the DL-reconstructed images exhibited a marked reduction in MAE (0.024 ± 0.002 vs. 0.076 ± 0.016 for ULF) and NMSE (0.035 ± 0.008 vs. 0.231 ± 0.047).

Table 1. Quantitative Evaluation Metrics for ULF, DL-reconstructed, and Reference MRI Images

Subject	Modality	SSIM (ULF/Ref)	SSIM (Rec/Ref)	MAE (ULF/Ref)	MAE (Rec/Ref)	PSNR (ULF/Ref)	PSNR (Rec/Ref)	NMSE (ULF/Ref)	NMSE (Re/Ref)
POCEMR003	T1-weighted	0.554	0.908	0.068	0.021	18.06	26.51	0.225	0.032
	T2-weighted	0.625	0.884	0.062	0.024	18.48	25.59	0.206	0.040
	FLAIR	0.501	0.887	0.099	0.025	15.15	25.28	0.245	0.024
POCEMR010	T1-weighted	0.535	0.885	0.076	0.027	17.22	24.84	0.209	0.036
	T2-weighted	0.635	0.875	0.049	0.023	20.28	25.99	0.147	0.040
	FLAIR	0.502	0.881	0.090	0.024	15.54	25.21	0.226	0.024
POCEMR024	T1-weighted	0.535	0.897	0.071	0.022	17.53	25.80	0.261	0.039
	T2-weighted	0.619	0.885	0.059	0.024	18.67	25.11	0.221	0.050
	FLAIR	0.493	0.880	0.088	0.027	15.78	24.33	0.229	0.032
POCEMR029	T1-weighted	0.537	0.891	0.077	0.023	17.08	25.62	0.291	0.041
	T2-weighted	0.593	0.873	0.059	0.025	18.8	25.23	0.206	0.047
	FLAIR	0.537	0.861	0.089	0.026	15.85	24.83	0.241	0.030
POCEMR088	T1-weighted	0.459	0.912	0.099	0.021	16.02	26.51	0.345	0.031
	T2-weighted	0.583	0.887	0.059	0.023	19.45	25.94	0.173	0.039
	FLAIR	0.4369	0.890	0.097	0.025	15.5	25.23	0.239	0.025

An examples of the ULF MR images, DL-reconstructed HF MR images, and the corresponding HF reference scans (Fig. 3). This visual comparison supports the quantitative findings and demonstrates that the method improves the quality of ULF MR images.

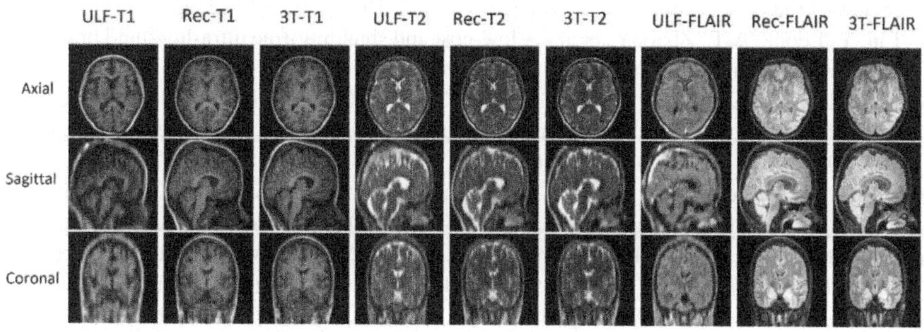

Fig. 3. Representative slices from T1-weighted, T2-weighted, and FLAIR MRI scans at ULF (64 mT), DL-reconstructed, and high field (3T). For each sequence and image type, axial, sagittal, and coronal views are provided.

For further evaluation on the unseen challenge data—where paired HF images were not provided—our model achieved a mean SSIM of 0.831, a mean PSNR of 22.843, a mean MAE of 0.032, and a mean NMSE of 0.032 across 10 images.

After submission and testing on 15 subjects (with background masking applied), the results showed SSIM of 0.714, PSNR of 30.319, MAE of 0.070, and NMSE of 0.068.

4 Discussion

The current work presents a GAN-based super-resolution framework to improve the resolution of ULF MRI images. The model is trained to learn the mapping of low-resolution 3D patches to their corresponding ground truth high-resolution patches. The results report significantly improved PSNR, SSIM, MAE, and NMSE. Finally, when evaluated on unseen challenge data, our super-resolved images demonstrated strong generalization and robustness of the proposed method.

One limitation of the current study may be related to the small sample size of the training, as suggested by the noticeable gap in SSIM performance between the validation and test sets. This indicates potential overfitting and suggests that a larger and more diverse dataset could improve the performance of the model. In future work, we will be extending this super-resolution approach to other ULF and HF paired MRI images from a wider range of clinical cases to test how the model performs on lesions and other abnormalities in the brain. In conclusion, the performance of our model across multiple MRI sequences demonstrates its robustness in improving image quality to help advance diagnosis and analytical applications.

Disclosure of Interests. The authors have no competing interests to declare that are relevant to the content of this article.

References

1. DeBenedectis, C.M., Spalluto, L.B., Americo, L., et al.: Health care disparities in radiology—a review of the current literature. J. Am. Coll. Radiol. **19**(1), 101–111 (2022)
2. Liu, Y., Leong, A.T., Zhao, Y., et al.: A low-cost and shielding-free ultra-low-field brain MRI scanner. Nat. Commun. **12**(1), 7238 (2021)
3. O'Reilly, T., Teeuwisse, W.M., De Gans, D., Koolstra, K., Webb, A.G.: In vivo 3D brain and extremity MRI at 50 mT using a permanent magnet Halbach array. Magn. Reson. Med. **85**(1), 495–505 (2021). https://doi.org/10.1002/mrm.28396
4. Arnold, T.C., Freeman, C.W., Litt, B., Stein, J.M.: Low-field MRI: clinical promise and challenges. J. Magn. Reson. Imaging **57**(1), 25–44 (2023). https://doi.org/10.1002/jmri.28408
5. Marques, J.P., Simonis, F.F.J., Webb, A.G.: Low-field MRI: an MR physics perspective. J. Magn. Reson. Imaging **49**(6), 1528–1542 (2019). https://doi.org/10.1002/jmri.26637
6. Trigka, M., Dritsas, E.: A comprehensive survey of deep learning approaches in image processing. Sensors **25**(2), 531 (2025)
7. Zhan, Z.H., Li, J.Y., Zhang, J.: Evolutionary deep learning: a survey. Neurocomputing **483**, 42–58 (2022)
8. Liu, Y., Zhang, Y., Wang, Y., et al.: A survey of visual transformers. IEEE Trans. Neural Netw. Learn. Syst. **35**(6), 7478–7498 (2023)

9. De Souza, V.L.T., Marques, B.A.D., Batagelo, H.C., Gois, J.P.: A review on generative adversarial networks for image generation. Comput. Graph. **114**, 13–25 (2023)
10. Croitoru, F.A., Hondru, V., Ionescu, R.T., Shah, M.: Diffusion models in vision: a survey. IEEE Trans. Pattern Anal. Mach. Intell. **45**(9), 10850–10869 (2023)
11. Chen, Y., Xie, Y., Zhou, Z., Shi, F., Christodoulou, A.G., Li, D.: Brain MRI super resolution using 3D deep densely connected neural networks. In: 2018 IEEE 15th International Symposium on Biomedical Imaging (ISBI 2018), pp. 739–742. IEEE (2018). https://ieeexplore.ieee.org/abstract/document/8363679/. Accessed 8 Aug 2025
12. Iglesias, J.E., Billot, B., Balbastre, Y., et al.: SynthSR: a public AI tool to turn heterogeneous clinical brain scans into high-resolution T1-weighted images for 3D morphometry. Sci. Adv. **9**(5), eadd3607 (2023). https://doi.org/10.1126/sciadv.add3607
13. Laguna, S., Schleicher, R., Billot, B., et al.: Super-resolution of portable low-field MRI in real scenarios: integration with denoising and domain adaptation. In: Medical Imaging with Deep Learning (2022). https://openreview.net/forum?id=pinw0Gcot4T&utm_medium=email&utm_source=transaction. Accessed 8 Aug 2025
14. Islam, K.T., Zhong, S., Zakavi, P., et al.: Improving portable low-field MRI image quality through image-to-image translation using paired low-and high-field images. Sci. Rep. **13**(1), 21183 (2023)
15. Hatamizadeh, A., Nath, V., Tang, Y., Yang, D., Roth, H.R., Xu, D.: Swin UNETR: swin transformers for semantic segmentation of brain tumors in MRI images. In: Crimi, A., Bakas, S. (eds.) Brainlesion: Glioma, Multiple Sclerosis, Stroke and Traumatic Brain Injuries. LNCS, vol. 12962, pp. 272–284. Springer International Publishing (2022). https://doi.org/10.1007/978-3-031-08999-2_22
16. Imambi, S., Prakash, K.B., Kanagachidambaresan, G.R.: PyTorch. In: Prakash, K.B., Kanagachidambaresan, G.R. (eds.) Programming with TensorFlow. EAI/Springer Innovations in Communication and Computing, pp. 87–104. Springer International Publishing (2021). https://doi.org/10.1007/978-3-030-57077-4_10
17. Cardoso, M.J., Li, W., Brown, R., et al.: MONAI: an open-source framework for deep learning in healthcare. arXiv. Preprint posted online 4 November 2022. https://doi.org/10.48550/arXiv.2211.02701

UltraMR-Enforce: A Unified Ensemble Framework for Enhancement of Ultra-Low-Field MRI

Xiaoyu Bai, Yueyue Zhu, Haotian Jiang, Rongqing Cai, Yi Liu, and Geng Chen(✉)

National Engineering Laboratory for Integrated Aero-Space-Ground-Ocean Big Data Application Technology, School of Computer Science and Engineering, Northwestern Polytechnical University, Xi'an, Shaanxi, China
geng.chen@ieee.org

Abstract. Magnetic Resonance Imaging (MRI) is a crucial non-invasive technique for visualizing brain anatomy. However, in resource-limited regions, the accessibility of high-field MRI scanners (e.g., 3T) is restricted due to equipment limitations, making it challenging to obtain high-quality images for clinical diagnosis. To address this issue, this study leverages paired high-field (3T) and ultra-low-field (64mT) MRI data to design and optimize various models, ultimately proposing an enhanced version of UltraMR-Enforce. The proposed model integrates the complementary strengths of Transformer and Mamba, further enhancing structural modeling and generalization capabilities through the weighted fusion of predictions from multiple models. Extensive experiments on the challenge dataset demonstrate that UltraMR-Enforce significantly improves the clarity and contrast of 64mT MRIs while precisely preserving critical anatomical structures, thereby further enhancing the clinical applicability and diagnostic reliability of ultra-low-field MR imaging.

Keywords: Ultra-Low-Field MRI · Enhancement Method · Deep Learning

1 Introduction

Magnetic Resonance Imaging (MRI) is a widely used non-invasive medical imaging technique that is extensively applied for observing brain structures and detecting pathological changes. High-field MRI (e.g., 3T) has become an essential tool in clinical practice due to its high resolution and excellent tissue contrast, enabling more accurate diagnosis and assessment of various neurological diseases [6]. However, high-field MRI equipment is typically expensive, bulky, and

---

X. Bai and Y. Zhu—Contributed equally to this work.

This work was supported in part by the National Natural Science Foundation of China under Grant 62571451 and the Innovation Capability Support Program of Shaanxi under Grant 2025ZC-KJXX-47.

Z. Chen et al. (Eds.): ULF-EnC 2025, LNCS 16293, pp. 84–93, 2026.
https://doi.org/10.1007/978-3-032-23344-8_10

requires demanding operational environments and skilled technicians, limiting its accessibility in resource-constrained and underdeveloped regions.

In contrast, Ultra-Low-Field (ULF) MRI (e.g., 64mT) offers advantages such as low cost, compact size, and ease of deployment, making it an ideal imaging solution for primary healthcare institutions and remote areas [1,10]. The widespread adoption of ULF MRI has the potential to significantly enhance medical diagnostic capabilities in these regions and promote healthcare equity. However, due to its inherently weak signal strength, ULF MRI suffers from low Signal-to-Noise Ratio (SNR) and poor tissue contrast, resulting in image quality far inferior to that of high-field MRI, which severely restricts its diagnostic accuracy and reliability [9]. Therefore, a critical challenge is how to significantly improve ULF MRI quality while preserving its device advantages.

In recent years, with the rapid development of deep learning technology, researchers have made remarkable progress in MRI reconstruction and enhancement. Early work by Wang et al. pioneered the use of Convolutional Neural Networks (CNNs) to map zero-filled images to high-quality MRIs, opening a new era of data-driven MRI reconstruction [13]. Subsequently, cascade CNN further optimized reconstruction effects through a multi-stage cascaded architecture, enhancing image detail representation [12]. Additionally, for MRI reconstruction across multiple anatomical regions, research teams have integrated various deep learning models to accelerate acquisition and improve image quality effectively [7,8,11,15]. Although these methods perform excellently in high-field MRI, their adaptability and effectiveness in the low SNR and complex imaging characteristics of ULF MRI remain insufficiently validated and limited.

To address this challenge, we propose an efficient and robust ULF MRI enhancement framework, named UltraMR-Enforce. This framework integrates multiple deep learning models to fully exploit their complementary strengths in feature extraction and image reconstruction, achieving precise enhancement of ULF MRIs. The specific contributions are as follows:

- We introduce a Transformer-based model that fully leverages its strength in global feature modeling to enhance image performance;
- We apply the emerging Mamba architecture to 3D ULF MRI tasks, effectively capturing long-range dependencies by leveraging three-dimensional information to further improve image quality.
- We integrate the predictions from multiple models through weighted fusion, which suppresses artifacts, enhances detail representation, and further improves the overall reconstruction quality.
- We conduct a systematic evaluation on the multi-modal paired brain MRI dataset from the ULF-EnC challenge, with results demonstrating that our method outperforms existing state-of-the-art approaches in both quantitative metrics and visual quality.

2 Method

2.1 Overall Architecture

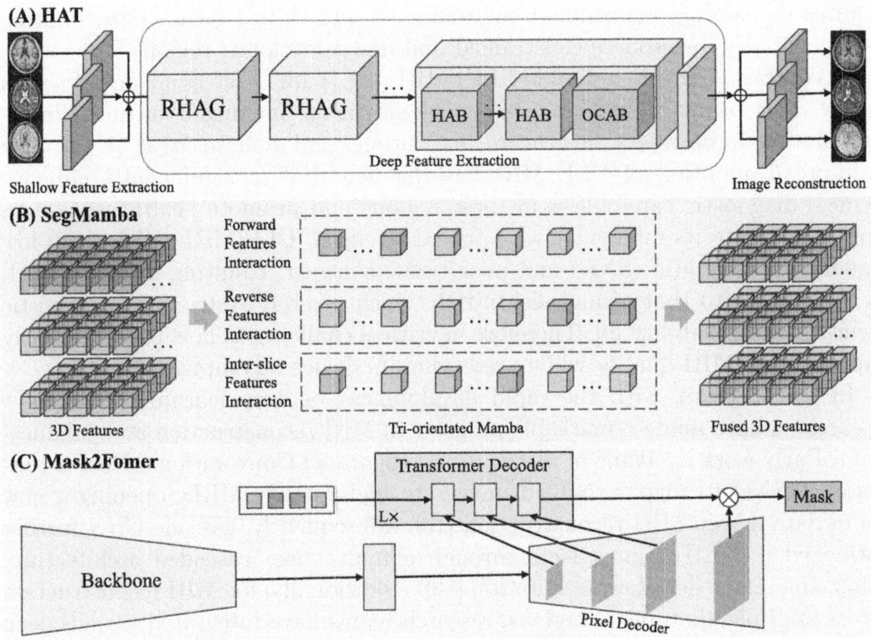

Fig. 1. Comparison of model architectures: (a) HAT: integrates Hybrid Attention Blocks (HAB) and Overlapping Cross-Attention Blocks (OCAB); (b) SegMamba: combines 3D information with Tri-oriented Mamba blocks for efficient feature extraction; (c) Mask2Former: utilizes multi-resolution features to enhance feature representation.

To effectively enhance the quality of ULF MRIs, we propose an ensemble-based enhancement framework, named UltraMR-Enforce. This method integrates three representative deep neural networks: Transformer-based HAT [4] and Mask2Former [5], both adept at capturing long-range global dependencies; and SegMamba [14], a Mamba-based architecture designed for efficient extraction of long-range 3D features from volumetric data. In addition, we further refine the enhancement results through weighted fusion. The following sections will detail the structural designs of these three sub-models and the weighted fusion method.

2.2 HAT

Hybrid Attention Block (HAB). HAB integrates Window-based Multi-head Self-Attention (W-MSA) and Channel Attention Mechanisms (CAB) to activate more pixel features and enhance the model's representation capability, as shown in Fig. 1 (A). The computation process can be expressed as:

$$X_M = \text{W-MSA}(X_N) + \alpha \cdot \text{CAB}(X_N) + X, \tag{1}$$

where X is the input feature, $X_N = \text{LN}(X)$ denotes the normalized intermediate feature, and α is a scaling factor. The window-based self-attention is computed within local windows of size $M \times M$, defined as:

$$\text{Attention}(Q, K, V) = \text{Softmax}\left(\frac{QK^\top}{\sqrt{d}} + B\right)V, \tag{2}$$

where Q, K, and V are the query, key, and value matrices, respectively, d is the dimension of the key vectors, and B represents the relative position encoding.

The CAB(·) consists of two convolutional layers and a channel attention module, designed to enhance feature representation by dynamically recalibrating channel-wise weights. By combining local spatial dependencies captured by self-attention with global channel-wise information, HAB(·) significantly improves the model's performance in image reconstruction tasks.

Overlapping Cross-Attention Block (OCAB). The OCAB enhances information interaction between neighboring windows by introducing an overlapping cross-window attention mechanism. Unlike the traditional non-overlapping window self-attention, OCAB partitions the query matrix X_Q of the input feature into non-overlapping windows of size $M \times M$, while the key matrix X_K and value matrix X_V are partitioned into overlapping windows of size $M_o \times M_o$. The overlapping window size is defined as:

$$M_o = (1 + \gamma) \times M, \tag{3}$$

where γ is a constant that controls the degree of overlap. This design allows each query window to access a larger range of contextual information, effectively mitigating boundary artifacts caused by traditional window partitioning.

2.3 SegMamba

The original Mamba block models global dependencies only along a single direction, which limits its ability to fully capture the complex spatial structures in high-dimensional medical images, making it unsuitable for volumetric feature modeling. To address this issue, SegMamba designs the Tri-orientated Mamba module within the TSMamba block, aiming to simultaneously compute and fuse feature dependencies from three different directions, thereby comprehensively capturing the global contextual information of high-dimensional features.

Specifically, as illustrated in Fig. 1 (B), SegMamba flattens the 3D input features into three one-dimensional sequences along the forward, reverse, and inter-slice directions. Each of these sequences is processed by a Mamba layer to model global dependencies and perform feature interactions. The outputs of the three sequences are then summed to produce the enhanced 3D feature representation:

$$\text{ToM}(\text{z}) = \text{Mamba}(\text{z}_f) + \text{Mamba}(\text{z}_r) + \text{Mamba}(\text{z}_s), \tag{4}$$

where Mamba(·) denotes the Mamba layer for modeling global dependencies in a sequence, and z_f, z_r, z_s represent the sequences flattened in the forward, reverse, and inter-slice directions, respectively. This multi-directional global modeling strategy effectively enhances the network's ability to understand the spatial structure of 3D medical images, significantly enriching and improving feature representation, which benefits downstream image analysis tasks.

2.4 Mask2Former

Mask2Former is a universal image processing architecture built upon a meta-architecture consisting of a feature extraction network, a pixel decoder, and a Transformer decoder, as shown in Fig. 1 (C). It introduces a core mechanism, masked attention, into the Transformer decoder, restricting the attention range to the foreground region of each predicted mask to focus on local feature extraction, accelerate convergence, and improve accuracy. The model adopts an efficient multi-scale high-resolution feature strategy, feeding features at different resolutions into the decoder layer by layer to enhance its ability to handle small targets. It also optimizes the decoder structure by reordering self-attention and cross-attention, introducing learnable query features, and removing dropout to improve information utilization.

During training, it calculates the mask loss on randomly sampled points, significantly reducing memory consumption while maintaining performance. With these designs, Mask2Former can achieve, and in some cases surpass, the performance of specialized models across different tasks using the same architecture, while offering higher training efficiency and better utilization of computational resources.

2.5 Weighted Fusion Strategy

We compute the weighted fusion coefficients based on the average PSNR of the three sub-models on the validation set. Let the average PSNR of the i-th model be PSNR_i, and let MAX denote the maximum pixel value. The corresponding Mean Squared Error (MSE) is given by:

$$\mathrm{MSE}_i = \frac{\mathrm{MAX}^2}{10^{\mathrm{PSNR}_i/10}}. \tag{5}$$

The fusion weight is then defined as the normalized reciprocal of the MSE:

$$w_i = \frac{1/\mathrm{MSE}_i}{\sum_{j=1}^{3} 1/\mathrm{MSE}_j}. \tag{6}$$

In this way, models with better performance are assigned higher weights. The computed weights are fixed for use on both the test and official datasets, fully leveraging the strengths of each model.

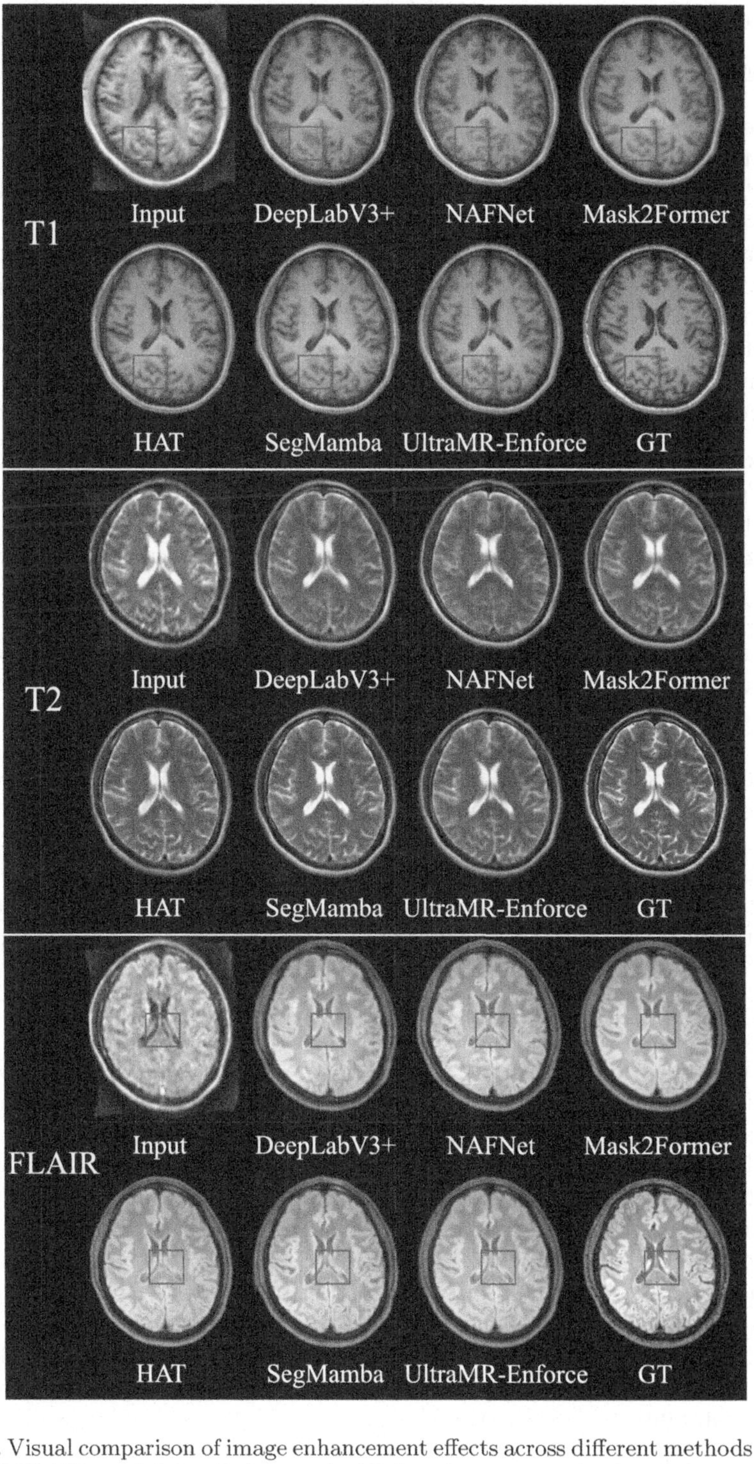

Fig. 2. Visual comparison of image enhancement effects across different methods on T1, T2, and FLAIR modalities. Red boxes highlight salient regions for detailed comparison. (Color figure online)

3 Results

3.1 Experimental Settings

Datasets. This study utilizes the paired brain MRI dataset provided by the ULF-EnC Challenge, which includes 64mT and 3T MRI scans from healthy participants. The data were acquired at Monash Biomedical Imaging, Monash University, Australia, using a Hyperfine Swoop scanner (64mT) and a Siemens Skyra scanner (3T). Experienced radiologists performed the scans following standardized protocols to ensure high quality and consistency. The dataset includes three common MRI sequences: T1-weighted, T2-weighted, and FLAIR, with paired acquisitions conducted under standardized conditions. The dataset consists of 50 training subjects, each with paired images across three modalities. To better evaluate the model's performance, 10 subjects were reserved as a validation set.

Implementation Details. All experiments are conducted on a computing platform equipped with two NVIDIA RTX 3090 GPUs. The model is implemented using the PyTorch deep learning framework. During training, we use the Adam optimizer and adopt the cosine annealing learning rate scheduler.

For the HAT-based model, the initial learning rate is set to 1e−4, and the batch size is set to two. During training, three consecutive slices along the D-dimension are selected as the input. The input has a resolution of $224 \times 224 \times 3$. All three modalities are used as input and are first passed through individual convolution layers before being concatenated together.

For the Mask2Former-based model, the initial learning rate is set to 1e−3, and the batch size is set to 64. The input images consist of three consecutive slices selected along the D-dimension.

For the SegMamba-based model, the initial learning rate is set to 1e-3, and the batch size is set to 64. The input images are divided into patches of size $32 \times 32 \times 32$. For each modality, one patch is extracted and concatenated across modalities as the final input.

3.2 Comparison with State-of-the-Art Models

Quantitative Results. To further verify the performance of the proposed method UltraMR-Enforce, we conducted comparative experiments with the current mainstream models DeepLabV3+ [2], NAFNet [3], Mask2Former [5], HAT [4], and SegMamba [14] under the same experimental environment and parameter settings. Table 1 presents the mean and standard deviation of Structural Similarity Index (SSIM), Peak Signal-to-Noise Ratio (PSNR), Normalized Mean Squared Error (NMSE), and Mean Absolute Error (MAE) for each method on the validation datasets. The results demonstrate that UltraMR-Enforce achieves the best overall performance across all metrics, outperforming the second-best method, SegMamba. Although it shows slightly higher variance in PSNR and NMSE, UltraMR-Enforce still exhibits superior overall image quality and reconstruction stability.

Table 1. Quantitative comparison of UltraMR-Enforce and other methods. The top of these results is shown in **bold**.

Model	SSIM	PSNR	MAE	NMSE	Final Score
DeepLabV3+	0.773 ± 0.025	21.845 ± **0.649**	0.037 ± 0.003	0.086 ± 0.013	2.913 ± **0.081**
NAFNet	0.717 ± 0.035	20.369 ± 0.686	0.044 ± 0.005	0.123 ± 0.017	2.722 ± 0.094
Mask2Former	0.793 ± 0.024	22.247 ± 0.689	0.035 ± 0.003	0.079 ± **0.012**	2.968 ± 0.085
HAT	0.802 ± 0.025	22.204 ± 0.884	0.035 ± 0.004	0.081 ± 0.017	2.971 ± 0.106
SegMamba	0.805 ± 0.036	22.646 ± 0.882	0.034 ± 0.004	0.073 ± 0.014	3.018 ± 0.107
UltraMR-Enforce	**0.815 ± 0.023**	**22.925 ± 0.810**	**0.032 ± 0.003**	**0.068 ± 0.013**	**3.053** ± 0.096

Visualization Analysis. Figure 2 clearly presents the qualitative analysis results of UltraMR-Enforce compared with other methods on a subject sample across T1, T2, and FLAIR modalities. Compared to existing methods, UltraMR-Enforce demonstrates significant advantages: UltraMR-Enforce consistently generates enhanced images with visually superior outcomes, exhibiting not only higher clarity but also more precise detail restoration. Particularly noteworthy is that UltraMR-Enforce achieves higher accuracy in preserving details and restoring boundary regions, while significantly reducing blurring effects and artifacts. This substantially improves the overall image quality and interpretability.

4 Discussion

In this study, we propose a ULF MRI enhancement method based on a multi-submodel ensemble, termed UltraMR-Enforce. This method effectively integrates the complementary strengths of multiple models and assigns fusion weights based on each submodel's performance on the validation set, allowing better-performing models to contribute more to the final prediction, thereby effectively suppressing biases and artifacts that may arise from individual models.

In future work, we plan to incorporate structural prior information, such as anatomical brain regions and gray-white matter distributions, to enhance the model's ability to identify and finely process key tissue areas. This will help maintain high-quality image reconstruction while improving the fidelity of diagnostically relevant features, further promoting the feasibility and practical value of ULF MRI in clinical applications.

5 Conclusion

To enhance the quality of ULF MRIs, this paper proposes an image enhancement method based on an ensemble of sub-models. The method leverages multiple submodels to learn image features from different perspectives, and optimizes their outputs through weighted fusion, achieving more accurate and robust image reconstruction. This approach not only fully exploits the complementary strengths of the sub-models in feature extraction and reconstruction but also effectively alleviates the low SNR commonly found in ULF MRIs.

Experimental results validate the effectiveness of the proposed method, showing that compared to single-model approaches, the ensemble architecture offers significant improvements in reconstruction quality and preservation of image details.

References

1. Arnold, T.C., Freeman, C.W., Litt, B., Stein, J.M.: Low-field MRI: clinical promise and challenges. J. Magn. Reson. Imaging **57**(1), 25–44 (2023)
2. Chen, L.C., Zhu, Y., Papandreou, G., Schroff, F., Adam, H.: Encoder-decoder with Atrous separable convolution for semantic image segmentation. In: Proceedings of the European Conference on Computer Vision (ECCV), pp. 801–818 (2018)
3. Chen, L., Chu, X., Zhang, X., Sun, J.: Simple baselines for image restoration. In: European Conference on Computer Vision, pp. 17–33. Springer (2022)
4. Chen, X., Wang, X., Zhou, J., Qiao, Y., Dong, C.: Activating more pixels in image super-resolution transformer. In: Proceedings of the IEEE/CVF Conference on Computer Vision and Pattern Recognition, pp. 22367–22377 (2023)
5. Cheng, B., Misra, I., Schwing, A.G., Kirillov, A., Girdhar, R.: Masked-attention mask transformer for universal image segmentation. In: Proceedings of the IEEE/CVF Conference on Computer Vision and Pattern Recognition, pp. 1290–1299 (2022)
6. Du, L., et al.: Unveiling the future: advancements in MRI imaging for neurodegenerative disorders. Ageing Res. Rev. **95**, 102230 (2024)
7. Johnson, P.M., Recht, M.P., Knoll, F.: Improving the speed of MRI with artificial intelligence. Semin. Musculoskelet. Radiol. **24**, 012–020 (2020)
8. Lin, D.J., Johnson, P.M., Knoll, F., Lui, Y.W.: Artificial intelligence for MR image reconstruction: an overview for clinicians. J. Magn. Reson. Imaging **53**(4), 1015–1028 (2021)
9. Liu, Y., et al.: A low-cost and shielding-free ultra-low-field brain MRI scanner. Nat. Commun. **12**(1), 7238 (2021)
10. Mazurek, M.H., et al.: Portable, bedside, low-field magnetic resonance imaging for evaluation of intracerebral hemorrhage. Nat. Commun. **12**(1), 5119 (2021)
11. Recht, M.P., et al.: Using deep learning to accelerate knee MRI at 3 T: results of an interchangeability study. Am. J. Roentgenol. **215**(6), 1421–1429 (2020)
12. Schlemper, J., Caballero, J., Hajnal, J.V., Price, A.N., Rueckert, D.: A deep cascade of convolutional neural networks for dynamic MR image reconstruction. IEEE Trans. Med. Imaging **37**(2), 491–503 (2017)
13. Wang, S., et al.: Accelerating magnetic resonance imaging via deep learning. In: 2016 IEEE 13th International Symposium on Biomedical Imaging (ISBI), pp. 514–517. IEEE (2016)

14. Xing, Z., Ye, T., Yang, Y., Liu, G., Zhu, L.: SegMamba: long-range sequential modeling mamba for 3D medical image segmentation. In: International Conference on Medical Image Computing and Computer-Assisted Intervention, pp. 578–588. Springer (2024)
15. Zhu, Y., Jiang, H., Cai, R., Chen, G.: Multi-label MambaOut for quality assessment of low-field pediatric brain MR images. In: MICCAI Challenge on LISA, pp. 3–11 (2024)

NAF-GAN: Anatomically Constrained GAN for Ultra-Low-Field MRI Enhancement

Zhenyu Xiang[1], Yicheng Wu[2(✉)], Ziyang Chen[1], Zaiyuan Liu[1], Yongsheng Pan[1], and Yong Xia[1]

[1] Northwestern Polytechnical University, Xi'an 710072, Shaanxi, China
[2] Monash University, Clayton 3168, VIC, Australia
yicheng.wu@monash.edu

Abstract. High-quality diagnostic imaging in resource-limited settings remains a significant challenge due to the limited accessibility of high-field Magnetic Resonance Imaging (MRI). The Ultra-Low-Field MRI Image Enhancement Challenge (ULF-EnC) addresses this issue by leveraging paired 64mT and 3T MRI scans to develop algorithms that enhance low-field image quality to high-field standards. In this paper, we propose an end-to-end NAF-GAN framework, a generative adversarial network based on the Nonlinear Activation Free Network (NAFNet) architecture, specifically designed for this task. Our model learns the complex translation from 64mT to 3T MRI sequences, improving image clarity and contrast while preserving the anatomical fidelity essential for clinical diagnosis. Both the patch-level adversarial loss and the segmentation-based Dice loss are designed to capture high-frequency details and preserve anatomical structures. The proposed approach achieves strong performance across T1, T2, and FLAIR contrasts, underscoring its potential to enable cost-effective and accessible MRI solutions for mobile and global healthcare applications.

Keywords: Ultra-Low-Field MRI · NAFNet · Anatomical Structures

1 Introduction

Magnetic Resonance Imaging (MRI) is a non-invasive medical imaging modality renowned for its exceptional soft-tissue contrast, making it indispensable for modern neuro-diagnostics [1]. However, the clinical standard of high-field MRI (e.g., 1.5T-3T) is hampered by significant limitations. These include high acquisition and maintenance costs, demanding infrastructure requirements, and inherent immobility, which necessitates transporting often critically ill patients to specialized facilities [21]. In response to these challenges, ultra-low-field (ULF) MRI systems, such as the 64mT scanner used in this challenge, have emerged as a promising alternative [16]. These systems offer affordability, portability, and reduced siting requirements, making them suitable for point-of-care applications

Z. Chen et al. (Eds.): ULF-EnC 2025, LNCS 16293, pp. 94–103, 2026.
https://doi.org/10.1007/978-3-032-23344-8_11

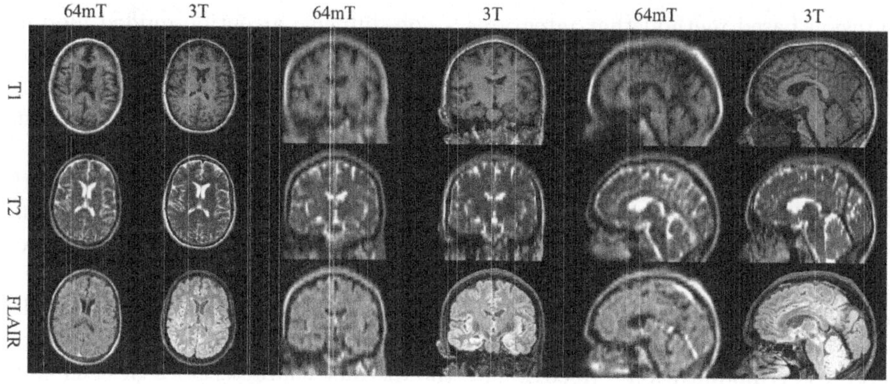

Fig. 1. Visual comparison of paired ultra-low-field (64mT) and high-field (3T) brain MRI scans across T1, T2, and FLAIR contrasts. The images highlight the significant difference in image quality and detail that the challenge aims to address.

in emergency departments, intensive care units, and resource-limited environments [2]. Despite these advantages, the clinical adoption of ULF MRI is fundamentally constrained by its inherently lower signal-to-noise ratio (SNR) and spatial resolution [15]. This results in image quality substantially inferior to the high-field standard, which not only hinders reliable diagnosis but also precludes the use of traditional quantitative analysis tools. Therefore, it is highly desirable to develop advanced image enhancement techniques to bridge this quality gap and unlock the full potential of accessible ULF MRI for global healthcare.

To accelerate innovation in this domain, the Ultra-Low-Field MRI Image Enhancement Challenge (ULF-EnC) was established. As shown in Fig. 1, this challenge focuses on enhancing 64mT images to approximate the quality of their 3T counterparts across multiple contrasts (T1, T2, and FLAIR). A core objective is to ensure that any enhancement not only improves visual clarity but also maintains the anatomical fidelity required for accurate clinical assessment of conditions such as stroke or neurological disorders.

Recent research has yielded several deep learning strategies for ULF to high-field MRI translation. Lucas et al. proposed LowGAN, a generative adversarial network composed of three parallel pix2pix architectures, each trained on a different orthogonal plane, which are subsequently combined using a wavelet-Fourier filter to form a final volume [17]. Hsu et al. [9] also utilized a GAN-based approach, introducing a CycleGAN framework trained on real ULF and high-field MRIs to improve the enhancement process. Kalluvila et al. [14] proposed a Nested U-Net (U-Net++) architecture that uses residual learning on synthetically downsampled data to reconstruct high-resolution images. Iglesias et al. [10] developed LF-SynthSR, which employs a 3D U-Net trained with a synthetic data generator for robustness and uniquely incorporates a segmentation head with an extra, frozen U-Net to guide the synthesis toward greater anatomical accuracy. More recently, Dayarathna et al. [5] introduced a novel adversarial diffusion-based method that integrates a non-diffusive attention-guided module to enhance informative features. Beyond proposing new architectures, Islam et al. [12] evaluated the effectiveness of existing models like SynthSR and LoHiResGAN in producing ULF-derived volumetric measurements that are consistent with those from 3T MRI.

In this paper, we propose NAF-GAN, a novel generative adversarial network specifically designed to address the ULF-EnC task by building upon the powerful NAFNet architecture. NAFNet (Nonlinear Activation Free Network) [4,22] is an efficient and effective image restoration baseline that forgoes complex nonlinear activations in favor of a simpler and more computationally efficient gating mechanism. To adapt this foundation for multi-modal MRI translation, we introduce several key modifications. Recognizing the distinct characteristics of each contrast, our generator features independent input and output heads for the T1, T2, and FLAIR modalities, which are connected by a shared deep backbone to learn common features. The adversarial framework is completed with a PatchGAN discriminator to enforce local realism in the synthesized images. Furthermore, to provide explicit anatomical guidance and preserve critical structures, we incorporate a segmentation-based Dice loss as an additional training constraint. We demonstrate that this integrated approach achieves a strong performance on this challenging task, effectively enhancing ULF images to a quality comparable to high-field standards.

2 Method

In this section, we describe the architecture of our proposed NAF-GAN (Fig. 2) and the objective function used for its training. Our proposed method, NAF-GAN, is a generative adversarial network designed to translate multi-modal ULF MRI to their high-field counterparts. The framework is composed of a multi-head generator, a PatchGAN discriminator, and an auxiliary Segmentator to provide anatomical guidance.

Generator. The core of our framework is a generator adapted from the NAFNet architecture [4], which is known for its efficiency and strong performance in image restoration tasks. A key innovation in NAFNet is the replacement of nonlinear activation functions like ReLU or GELU with a SimpleGate mechanism. This mechanism splits the feature channels into two halves and performs an element-wise multiplication, an operation that reduces computational complexity while proving highly effective. We adapt this efficient architecture with a multi-head, shared-backbone design. The process begins with three independent input heads—one for each modality (T1, T2, and FLAIR)—where each head consists of a 2D convolution followed by a NAFBlock to generate a 32-channel feature map. These features are then concatenated along the channel axis, and a single NAFBlock is used to compress them into a 32-channel representation before they are fed into the shared backbone. The backbone itself has a depth of 4 levels, with each level employing a single NAFBlock for processing. Finally,

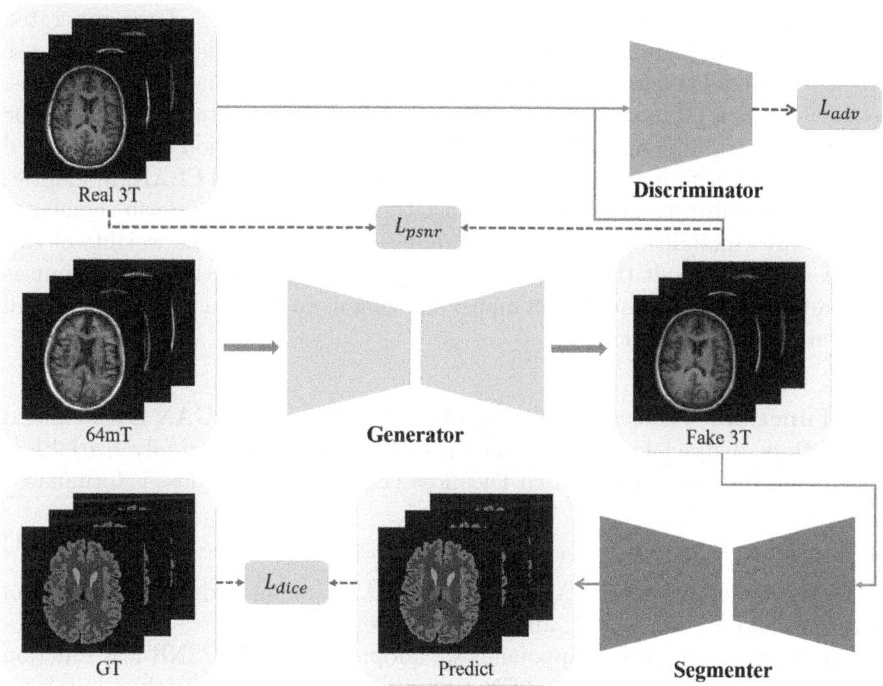

Fig. 2. Architecture of our NAF-GAN framework for ultra-low-field to high-field MRI translation. There are three main components. **(1) Generator:** Adapted from the NAFNet architecture, the generator takes multi-contrast low-field (64mT) images as input. **(2) Discriminator:** A PatchGAN discriminator provides an adversarial loss by differentiating between the generated images and the ground truth 3T scans, enforcing local textural realism. **(3) Segmenter:** To ensure anatomical fidelity, the synthesized 3T image is fed into an extra ResUNet Segmenter with frozen weights.

the 32-channel features from the backbone are passed to three independent 2D convolutional output heads, each projecting the features back to a single channel to reconstruct the final high-field modalities.

Discriminator. For the adversarial component, we utilize a PatchGAN discriminator [13]. Our implementation takes a 3-channel input, corresponding to the concatenated T1, T2, and FLAIR modalities. The network consists of 3 convolutional layers, where the first layer maps the input to 32 feature channels. This PatchGAN architecture operates on local image patches, classifying each N×N patch as real or fake. This design choice encourages the generator to produce high-frequency details and locally realistic textures, which are crucial for preserving the fine anatomical structures present in high-quality medical images.

Segmenter. To further improve the anatomical correctness of the generated images, we incorporate an auxiliary Segmenter as an anatomical expert. We use a ResUNet model [19] for this purpose. Instead of relying on manual annotations, we use SynthSeg [3] to generate pseudo-ground truth segmentation maps for the high-field 3T training data. We observed that the segmentation maps produced by SynthSeg exhibit notable differences across the T1, T2, and FLAIR modalities for the same subject. Therefore, our ResUNet is trained to perform segmentation on three modalities. During the training of NAF-GAN, the weights of this ResUNet are frozen. It takes the generator's synthesized multi-modal output as input and produces segmentation maps, allowing us to apply a direct anatomical constraint via the Dice loss.

Loss Function. The overall training objective of our NAF-GAN is a weighted sum of three distinct loss components: a PSNR loss ($\mathcal{L}_{\text{psnr}}$), an adversarial loss ($\mathcal{L}_{\text{adv}}$), and a segmentation-based Dice loss ($\mathcal{L}_{\text{dice}}$). The total loss is formulated as:

$$\mathcal{L}_{\text{total}} = \lambda_{\text{psnr}}\mathcal{L}_{\text{psnr}} + \lambda_{\text{adv}}\mathcal{L}_{\text{adv}} + \lambda_{\text{dice}}\mathcal{L}_{\text{dice}} \tag{1}$$

where the hyperparameters λ_{psnr}, λ_{adv}, and λ_{dice} balance the contribution of each term and are set to 2, 1, and 1, respectively.

For the reconstruction component, we adopt the specific PSNR loss function as implemented in the official NAFNet paper [4]. Instead of directly minimizing the Mean Squared Error (MSE), this loss minimizes a scaled logarithm of the MSE. This objective is functionally equivalent to directly maximizing the PSNR metric. For a generated image $G(x)$ and a ground truth image y, the loss is defined as:

$$\mathcal{L}_{\text{psnr}} = \mathbb{E}_{x,y}\left[\frac{10}{\log(10)}\log\left(\|y - G(x)\|_2^2 + \epsilon\right)\right] \tag{2}$$

where $\|\cdot\|_2^2$ denotes the mean squared error calculated over the image dimensions, and ϵ is a small constant set to 1×10^{-8} for numerical stability.

We employ the Hinge GAN loss [8] and apply an R1 gradient penalty [18] to the discriminator for training stability. In this framework, the generator G maps a low-field input x to a synthetic image $G(x)$. The discriminator D is trained to distinguish these generated images from real high-field images y. The adversarial loss is formulated as:

$$\mathcal{L}_{\text{adv}} = \mathbb{E}_y[\max(0, 1 - D(y))] + \mathbb{E}_x[\max(0, 1 + D(G(x)))] \tag{3}$$

To explicitly enforce structural fidelity, we compute a Dice loss using the extra ResUNet Segmenter, S. The generated image $G(x)$ is passed through the frozen Segmenter, and the resulting segmentation map $S(G(x))$ is compared against the ground truth segmentation map y_{gt}. The Dice loss [23] is defined as:

$$\mathcal{L}_{\text{dice}} = 1 - \frac{2|S(G(x)) \cap y_{\text{gt}}|}{|S(G(x))| + |y_{\text{gt}}|} \tag{4}$$

3 Experiment and Result

Datasets. The dataset used in this study is the official release from the ULF-EnC Challenge [5–7,11]. It consists of paired brain MRI scans acquired on a Siemens Biograph mMR (3T) scanner and a Hyperfine Swoop (64mT) scanner. The dataset is divided into a training set with 50 subjects and a validation set with 10 subjects, with T1-weighted, T2-weighted, and FLAIR modalities provided for each. The challenge organizers performed an official pre-processing pipeline, which included key steps such as bias field correction using FSL-FAST and the co-registration of the 64mT and 3T scans (Size: $224 \times 224 \times 160$, Isotropic spacing: $1 \times 1 \times 1$mm and Intensity values $\in [0, 255]$).

Implementation Details. To address the issue of incomplete 64mT slices and improve inter-slice continuity, we adopted a training strategy that involves simultaneously loading five consecutive axial slices as input to predict the central slice [20]. For the training process, we extracted these 5-slice stacks, with dimensions of $224 \times 224 \times 5$, from the volumetric images of all three modalities (T1, T2, and FLAIR). The data was linearly scaled to the range [0, 1]. To enhance model robustness, we applied data augmentation to each image stack, which included: (1) random rotations with an angle uniformly sampled from $[-10, 10]^{\circ}$, (2) random elastic deformations defined by a coarse grid of 7×7 control points with locked image borders, and (3) RandomErasing. The model was trained for 600 epochs with a batch size of 4. We utilized the Adam optimizer with a fixed learning rate of 1×10^{-3} and parameters ($\beta_1 = 0.5, \beta_2 = 0.9$). All experiments were conducted on a single NVIDIA GeForce 4070Ti SUPER GPU.

To incorporate anatomical prior knowledge, we first used SynthSeg [3] to generate pseudo-ground truth segmentation maps from the high-field data. We refined these maps by mapping the left and right cerebral hemisphere labels to a single consistent set and addressing inconsistencies observed for the inferior lateral ventricle and CSF regions. Specifically, the inferior lateral ventricle was merged with the lateral ventricle label, and CSF was mapped to the background, resulting in a final set of 16 labels (indexed 0–15). These refined maps were then used to train our auxiliary Segmenter, for which we adopted a ResUNet model. The Segmenter was trained for 500 epochs using a combined Cross-Entropy and Dice loss, a learning rate of 1×10^{-4}, and a batch size of 4.

We evaluated our proposed NAF-GAN on the official ULF-EnC validation set. The quantitative results of our approach, benchmarked against other competing methods, are presented in Table 1. Our final model achieved an SSIM of 0.709, a PSNR of 29.637 dB, an MAE of 0.068, and an NMSE of 0.146. These results demonstrate that our method remains highly competitive. Furthermore, we conducted the ablation study in Table 1. First, the introduction of the anatomical constraint via the Dice loss yielded a clear improvement across all metrics, confirming its effectiveness in enforcing structural fidelity. Building upon this, we further explored a 2.5D multi-slice input strategy. The results show a consistent and significant performance gain as we increased the number of input slices from one to three, and subsequently to five. Our final 5-slice model

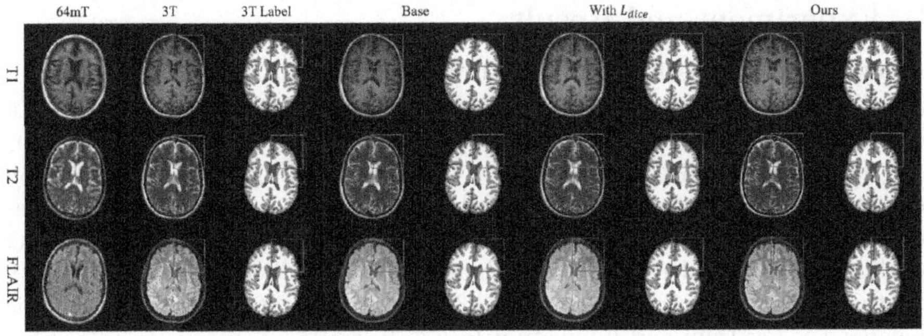

Fig. 3. Generation of 3T images from 64mT inputs.

Table 1. Ablation study of the proposed method. An improvement in performance is observed after introducing the segmentation Dice loss and a multi-slice input strategy.

METHOD	SSIM	PSNR	MAE	NMSE	FINAL SCORE
Baseline	0.667	28.625	0.076	0.175	3.504
+ $\mathcal{L}_{\text{dice}}$	0.675	28.870	0.073	0.164	3.536
+ 3 Slices	0.695	29.328	0.071	0.161	3.596
+ 5 Slices (Ours)	**0.709**	**29.637**	**0.068**	**0.146**	**3.639**
Testing (Ranking #5)	0.711	29.514	0.068	0.063	0.777

achieved the best performance in our experiments, underscoring the benefits of leveraging inter-slice context for more robust image synthesis.

As highlighted by the red box in Fig. 3, a visual comparison reveals that the inclusion of the Dice loss leads to the generation of images with sharper and more accurate anatomical details, particularly in the delineation of tissue boundaries. The multi-slice strategy further refines these outputs, resulting in images that more closely resemble the high-field ground truth. This visual evidence corroborates our quantitative findings and highlights the model's improved ability to reconstruct clinically relevant structures with high fidelity.

During our experiments, we identified several data quality challenges within the official ULF-EnC dataset. Beyond imperfect spatial registration between paired scans, we noted significant instances of incomplete brain tissue within certain axial slices, a problem clearly exemplified in subject POCEMR027 from the validation set. This data scarcity at the slice level poses a fundamental challenge for any 2D image synthesis method, as it requires the model to hallucinate missing anatomical structures from a deficient input. To address this issue and improve inter-slice continuity, we experimented with a training strategy that involves simultaneously loading multiple consecutive axial slices as input to predict the central slice [20]. The rationale is that the model can leverage contextual information from adjacent, more complete slices to better reconstruct the tar-

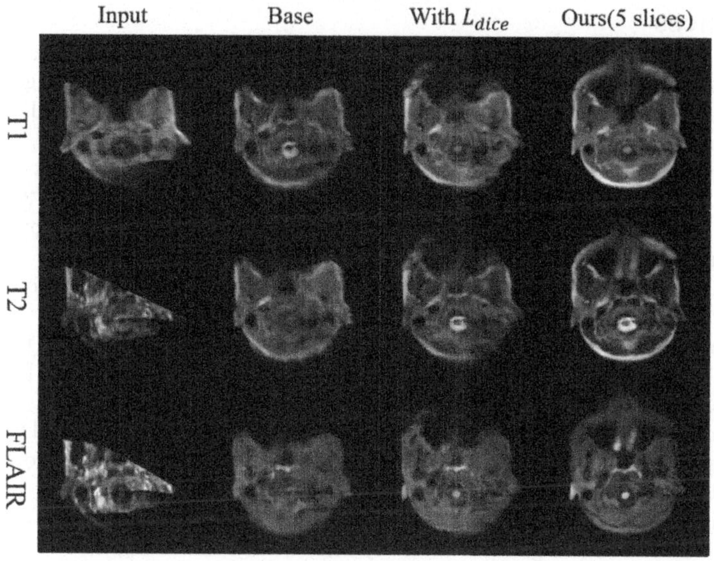

Fig. 4. Qualitative comparison of different strategies for handling an incomplete input slice from subject POCEMR027. From left to right: (a) The original 64mT input. (b) Output from the baseline model. (c) Output from the model with the $\mathcal{L}_{\text{dice}}$ loss. (d) Output from our experimental model trained with a 5-slice input.

get. When faced with the incomplete input slice, both the baseline model and the model with Dice loss struggle to generate a plausible anatomical structure (See Fig. 4). In contrast, the model trained with the multi-slice input strategy successfully alleviates this issue. It generates a more structurally complete brain image, demonstrating that providing local slice context is an effective approach for making models more robust to these types.

4 Conclusion

We have presented NAF-GAN, a generative adversarial network tailored for the Ultra-Low-Field MRI Image Enhancement Challenge. Our approach adapted the efficient NAFNet architecture with a multi-head design to handle multi-modal inputs and, crucially, incorporated an anatomical constraint via a segmentation-based Dice loss. Our method achieved a competitive quantitative score, excelling in structural similarity, which was a primary focus of the challenge. Our ablation studies definitively confirmed that the addition of the Dice loss was instrumental in improving anatomical accuracy and overall performance. These results contribute to the ongoing effort to make ULF MRI a more reliable and accessible diagnostic tool. Future work will focus on improving model robustness to various data degradations, such as the registration and slice-completeness issues, to further enhance its clinical applicability.

References

1. Abate, F., et al.: Unity: a low-field magnetic resonance neuroimaging initiative to characterize neurodevelopment in low and middle-income settings. Dev. Cogn. Neurosci. **69**, 101397 (2024)
2. Altaf, A., et al.: Applications, limitations and advancements of ultra-low-field magnetic resonance imaging: a scoping review. Surg. Neurol. Int. **15**, 218 (2024)
3. Billot, B., et al.: Synthseg: segmentation of brain mri scans of any contrast and resolution without retraining. Med. Image Anal. **86**, 102789 (2023)
4. Chen, L., Chu, X., Zhang, X., Sun, J.: Simple baselines for image restoration. In: European Conference on Computer Vision, pp. 17–33. Springer (2022)
5. Dayarathna, S., Islam, K.T., Chen, Z.: Ultra low-field to high-field mri translation using adversarial diffusion. In: 2024 IEEE International Symposium on Biomedical Imaging (ISBI), pp. 1–4. IEEE (2024)
6. Dayarathna, S., Islam, K.T., Uribe, S., Yang, G., Hayat, M., Chen, Z.: Deep learning based synthesis of mri, ct and pet: review and analysis. Med. Image Anal. **92**, 103046 (2024)
7. Dayarathna, S., et al.: Mccad: multi-contrast mri conditioned, adaptive adversarial diffusion model for high-fidelity mri synthesis. In: 2025 IEEE/CVF Winter Conference on Applications of Computer Vision (WACV), pp. 670–679. IEEE (2025)
8. Goni, O., Arka, H.S., Halder, M., Shibly, M.M.A., Shatabda, S.: Hingerlc-gan: combatting mode collapse with hinge loss and rlc regularization. In: International Conference on Pattern Recognition, pp. 370–385. Springer (2024)
9. Hsu, P., Marchetto, E., Sodickson, D., Johnson, P., Veraart, J.: Advancing the reliability of ultra-low field mri brain volume analysis using cyclegan. In: Meets Africa Workshop, pp. 52–62. Springer (2024)
10. Iglesias, J.E., et al.: Quantitative brain morphometry of portable low-field-strength mri using super-resolution machine learning. Radiology **306**(3), e220522 (2022)
11. Islam, K.T., et al.: Improving portable low-field mri image quality through image-to-image translation using paired low-and high-field images. Sci. Rep. **13**(1), 21183 (2023)
12. Islam, K.T., et al.: Ai improves consistency in regional brain volumes measured in ultra-low-field mri and 3t mri. Front. Neuroimaging **4**, 1588487 (2025)
13. Isola, P., Zhu, J.Y., Zhou, T., Efros, A.A.: Image-to-image translation with conditional adversarial networks. In: Proceedings of the IEEE Conference on Computer Vision and Pattern Recognition, pp. 1125–1134 (2017)
14. Kalluvila, A., Koonjoo, N., Bhutto, D., Rockenbach, M., Rosen, M.S.: Synthetic low-field mri super-resolution via nested u-net architecture. arXiv preprint arXiv:2211.15047 (2022)
15. Lau, V., et al.: Pushing the limits of low-cost ultra-low-field mri by dual-acquisition deep learning 3d superresolution. Magn. Reson. Med. **90**(2), 400–416 (2023)
16. Lim, T.R., et al.: Low-field (64 mt) portable mri for rapid point-of-care diagnosis of dissemination in space in patients presenting with optic neuritis. Am. J. Neuroradiol. **45**(11), 1819–1825 (2024)
17. Lucas, A., et al.: Multisequence 3-t image synthesis from 64-mt low-field-strength mri using generative adversarial networks in multiple sclerosis. Radiology **315**(1), e233529 (2025)
18. Mescheder, L., Geiger, A., Nowozin, S.: Which training methods for gans do actually converge? In: International Conference on Machine Learning, pp. 3481–3490. PMLR (2018)

19. Pan, Y., Ye, Y., Zhang, Y., Xia, Y., Shen, D.: Draw sketch, draw flesh: whole-body computed tomography from any x-ray views. Int. J. Comput. Vision **133**(5), 2505–2526 (2025)
20. Qu, L., Zhang, Y., Wang, S., Yap, P.T., Shen, D.: Synthesized 7t mri from 3t mri via deep learning in spatial and wavelet domains. Med. Image Anal. **62**, 101663 (2020)
21. Shoghli, A., Chow, D., Kuoy, E., Yaghmai, V.: Current role of portable mri in diagnosis of acute neurological conditions. Front. Neurol. **14**, 1255858 (2023)
22. Wu, Y., et al.: Codebrain: imputing any brain mri via modality-and instance-specific codes. arXiv preprint arXiv:2501.18328 (2025)
23. Zhao, R., et al.: Rethinking dice loss for medical image segmentation. In: 2020 IEEE International Conference on Data Mining (ICDM), pp. 851–860. IEEE (2020)

Enhancing Ultra-Low-Field MRI with Segmentation-Guided Adversarial Learning

James Grover(✉), Andrew Phair, Michael Ferraro, and David E. J. Waddington

Image X Institute, Sydney School of Health Sciences, Faculty of Medicine and Health, The University of Sydney, Sydney, Australia
james.grover@sydney.edu.au

Abstract. Ultra-low-field (ULF) MRI offers portable and low-cost imaging but suffers from poor image quality. To address this, we present our submission to the 2025 ULF Enhancement Challenge (ULF-EnC), where the goal is to synthesise high-field-like images from 64 mT scans. Our pipeline enhances ULF images through a combination of anatomical conditioning and model ensembling. We first generate tissue segmentation priors using a Swin UNETR trained solely on challenge-provided data. These priors condition two independent enhancement networks, a CycleGAN and a transformer-based residual enhancement model (T-REX), each trained to synthesise 3 T-like images. Outputs from both models are combined using a weighted average. Our approach produces enhanced images that were comparable to high-field scans both quantitatively and qualitatively.

Keywords: ULF-EnC Challenge · Ultra-low-field MRI · Image synthesis

1 Introduction

Magnetic resonance imaging (MRI) provides unparalleled structural and functional insight into human anatomy, yet remains inaccessible to much of the global population due to two major limitations: cost and portability. High-field (HF) MRI systems (e.g., 3 T) are expensive, stationary, and require specialised infrastructure, making them largely unavailable in low-and middle-income countries and impractical for critically ill or immobile patients [15]. In response, portable ultra-low-field (ULF) MRI has emerged as a promising alternative, enabling bedside and community-based imaging with lower infrastructure demands [20,26]. These systems operate at much lower magnetic field strengths (e.g., 64 mT), leading to significantly reduced signal-to-noise ratio (SNR), suboptimal image contrast, and susceptibility to artefacts from limited RF shielding and gradient performance. Addressing these limitations to unlock the diagnostic potential of ULF MRI remains a critical challenge for equitable medical imaging.

Z. Chen et al. (Eds.): ULF-EnC 2025, LNCS 16293, pp. 104–114, 2026.
https://doi.org/10.1007/978-3-032-23344-8_12

Recent research has focused on applying machine learning to enhance ULF MRI by translating low-SNR ULF scans into HF-quality images. Most of these approaches leverage paired datasets with subjects that have been imaged with both ULF and conventional HF MRI, enabling supervised learning and evaluation. Since radiologists and clinicians are already familiar with the contrast and resolution characteristics of 3-T images, synthesising images in this domain can aid interpretation and increase clinical confidence. The Ultra-Low-Field MRI Image Enhancement Challenge (ULF-EnC) [25] consists of paired 64-mT and 3-T brain MR images across multiple contrasts (T1, T2, FLAIR).

Paired image-translation approaches have demonstrated that generative models can lift 64-mT images toward 3 T appearance, improving visual quality and enabling standard neuro-morphometry analyses [6,9,10,18]. However, existing enhancement pipelines rarely condition synthesis on explicit tissue logits or combine complementary generators within the stringent data constraints of portable ULF MRI—limitations that the present study seeks to overcome.

Here we demonstrate a structured deep learning pipeline for enhancing ULF (64 mT) images using paired data from the ULF-EnC challenge. Our approach is guided by the principle that any successful enhancement method must ultimately infer and preserve underlying brain anatomy. To this end, we adopt a segmentation-driven strategy: anatomical priors in the form of logits are extracted via Swin UNETR trained to predict HF tissue maps generated using automated FSL tools (i.e. using no external data). The resulting logits are used to condition two complementary enhancement networks—a CycleGAN and a transformer-based residual enhancement model (T-REX). Their outputs are combined using a weighted average to produce a final enhancement. We hypothesised that embedding anatomical structure into the enhancement process is essential for diagnostic reliability. Additionally, by combining the enhancements from independently trained models image quality and robustness could be increased over single model approaches.

2 Methods

2.1 Dataset

The ULF-EnC training dataset is comprised of 50 subjects, each containing ULF (64 mT, *Hyperfine* Swoop) and HF (3 T, *Siemens* Biograph mMR) brain MRIs with T1, T2, and FLAIR contrast [2–4,10].

2.2 Overall Pipeline

We developed a pipeline for enhancing 64 mT ULF multi-contrast inputs (Fig. 1). The pipeline consists of four phases:

1. Segmentation prediction using Swin UNETR,
2. ULF enhancement using CycleGAN,
3. ULF enhancement using T-REX, and

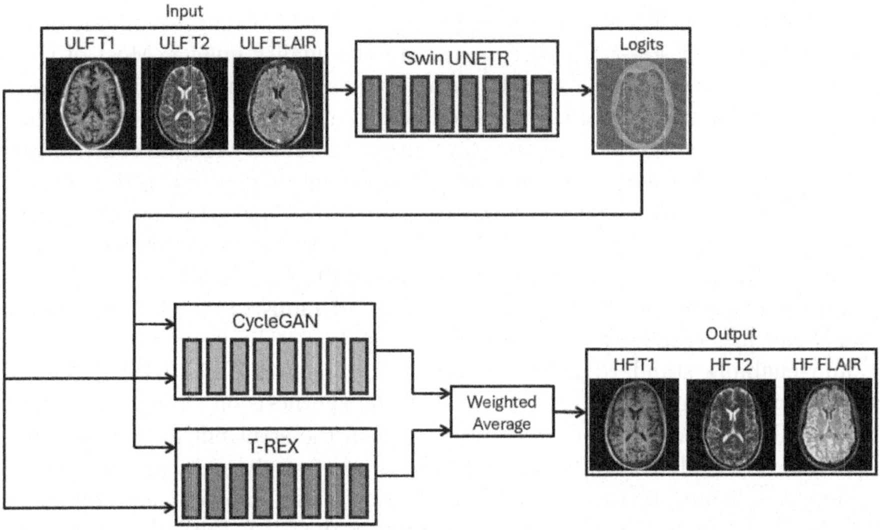

Fig. 1. Overall pipeline for producing enhanced ultra-low-field images. Input ultra-low-field images are enhanced by individual CycleGAN and T-REX networks with additional segmentation priors by Swin UNETR. Enhancements are combined to produce final output images.

4. Enhancement combination.

All models were trained and implemented using the PyTorch deep learning framework [22].

2.3 Segmentation Prediction Using Swin UNETR

Pre-trained neural networks can improve the performance of downstream neural networks by injecting additional priors. We hypothesised that by training a segmentation network to generate segmentations of various anatomical structures from the ULF images, deep features could be transferred to the task of enhancement.

To generate ground truth segmentations for training, we developed a multi-step preprocessing pipeline applied to the 3 T T1- and T2-weighted images. Importantly, caution was taken to ensure that no external data or pre-trained models were used at any stage of this segmentation process—only the challenge-provided data and publicly available, non-learning-based and non-atlas-based tools were used.

FSL's Brain Extraction Tool (BET) was applied to the 3 T T1- and T2-weighted images jointly, producing brain, skull, and scalp masks [12,13,24]. BET employs a deformable surface model initialised as a sphere, which expands iteratively to fit the brain boundary by optimising intensity contrast and surface smoothness.

FSL's FAST was then used to segment the T1-weighted brain image into cerebrospinal fluid (CSF), grey matter (GM), and white matter (WM) [28]. FAST uses a hidden Markov random field (HMRF) model with expectation-maximisation to simultaneously perform bias field correction and tissue classification.

The resulting masks were combined into a six-class labelmap: background, CSF, GM, WM, skull, and scalp. A Swin UNETR model was trained to predict the six-class 3 T segmentation directly from the three 64 mT contrasts (T1, T2, and FLAIR) [7]. The network architecture used 3D Swin Transformer blocks with hierarchical encoding, and was implemented using the MONAI framework [1,17]. The number of trainable parameters of this model was $\approx$ 62M.

The loss function was a combination of multi-class Dice loss and cross-entropy loss. Data augmentation was applied using MONAI's 3D transform library, including random affine transformations, flipping, and intensity perturbations. In the final training phase, augmentation was disabled and training continued on unaltered data. Model performance was evaluated using the mean Dice similarity coefficient on a separate validation set. Segmentation model weights were frozen after this training for all subsequent use.

2.4 ULF Enhancement Using CycleGAN

The image-to-image translation framework CycleGAN was adapted to incorporate 3D network architectures and trained to perform the task of low-field-to-high-field enhancement across three image contrasts simultaneously [29].

The principle underlying the CycleGAN approach is that by simultaneously training two generator-discriminator network pairs operating in opposite directions, the conditional generators learn to generate images that belong within the target output distribution while also relating to the input image. To enforce this, cycle losses in each domain are calculated between input images and the output that is yielded by applying the two generators successively.

The architecture of the generator network [8] was adapted from the 9-block residual network included in the publicly available CycleGAN code. We note that this 2D generator architecture has previously been successfully applied to an ULF MRI enhancement task [10]. To provide greater volumetric context, 2D convolution operations were replaced with their 3D equivalent and, to reduce chequerboard artefacts, transpose convolution operations were replaced with trilinear upsampling followed by 3D convolutions.

To condition our generator with segmentation information we employed spatially adaptive normalisation (SPADE) to enhance the inductive bias provided by semantic conditioning, replacing all normalisation layers in the ULF-to-HF generator with SPADE [21]. The final generator architecture had $\approx$ 33M parameters.

The HF-to-ULF generator received only the three-channel HF image volumes as input, and output corresponding synthesised three-channel low-field image volumes. The two corresponding discriminator networks were trained to distinguish between genuine and generated images in each domain.

An additional paired loss term was introduced between the HF and enhanced low-field 3D image slabs, calculated as the negative of the weighted challenge score (1). Each CycleGAN network was trained for a total of 500 epochs using the Adam optimiser [14], with an initial learning rate of 2×10^{-4} maintained for 100 epochs, followed by a linear decay to 0 over the remaining 400 epochs. The existing CycleGAN loss functions were scaled using the default penalty weights [29], while the paired loss term was weighted by a heuristically selected function given as the arctangent of the epoch number, scaled to take values between 0 and 80 over the course of the training. The intention behind the latter was to force the generator to first learn to produce enhanced images consistent with the HF image distribution, and then to fine-tune these enhanced images to optimise the challenge metrics.

Due to both memory and data scarcity constraints, the CycleGAN was trained on 40-slice slabs randomly selected from the total 3D image volumes each epoch, with the batch size also being limited to 1. At inference, overlapping slabs (with a stride of 5) were enhanced and then averaged to form a composite enhanced volume.

2.5 ULF Enhancement Using T-REX

We developed a 3D volume-to-volume transformer residual hybrid network (T-REX) for image synthesis. T-REX is comprised of an encoding arm, decoding arm, and information bottleneck utilising skip connections between encoder and decoder (similar to a U-NET) [23]. The encoding arm consists of convolutional layers and downsampling layers via strided convolution. The decoding arm consists of convolutional layers and upsampling layers via interpolation. At the information bottleneck, the latent tensor is transformer to a linear representation (length 512) and processed by a transformer encoder (two layers and four attention heads) [27]. This processed latent tensor is then reshaped back into the image domain and provided to the decoding arm. Skip connections between stages of the encoding and decoding arm incorporate residual blocks based off the enhanced deep super-resolution (EDSR) architecture [16]. The number of residual blocks are 16, 8, 4, and 2 based on the downsampling stage of the U-NET. Segmentation information (produced by the pre-trained Swin UNETR) were concatenated as six additional channels to the ULF images. The number of trainable parameters for T-REX was $\approx$ 24M.

Due to the significant distribution shift between ULF and HF, T-REX was trained adversarially using a conditional GAN (cGAN) framework [11]. The generator loss included two terms, a discriminator term (based on least squares) and a content term [19]. The content term ensures that generated enhancements closely resemble the HF MRIs on a pixel-wise basis; This was a weighted combination of L1 loss, PSNR loss, and a Sobel filter-based loss (to directly optimise for high spatial frequency features). T-REX was trained using an Adam optimiser [14], a batch size of 1, and a learning rate of 1×10^{-4} for both the generator and discriminator. The generator learning rate was reduced by a factor of $\frac{2}{3}$ at every quarter of training. Once T-REX had converged under adversarial training ($\approx$ 100 epochs), the generator was fine-tuned in isolation ($\approx$ 10 epochs).

2.6 Enhancement Combination

Model ensembling is a powerful technique to ultimately boost performance and increase robustness in the context of medical imaging deep learning [5]. For both the CycleGAN and T-REX, we implemented cross-validation training to produce a series of trained parameters for each model (six for CycleGAN and seven for T-REX). When performing inference, each model's final prediction was the average over the model's corresponding cross-validation folds. We then combined the final CycleGAN and T-REX predictions using a simple weighted average (70:30 CycleGAN:T-REX). In this way, a single enhancement could be produced from two independently trained networks over several cross-validation folds. We optimised this weighting on the held-out ULF-EnC validation set.

2.7 Measuring Performance

A qualitative and quantitative analysis was performed on the enhanced validation ULF images from the ULF EnC challenge. Qualitative analysis included the inspection of enhancements with regards to overall image quality, hallucinated features, and artefacts. We compare each individual model in addition to the combination from two cases of an internal held-out validation set.

We use the quantitative metrics as measured by challenge organisers to ensure consistent reporting with the final hidden test set. Two sets of metrics were reported: one involving no use of masking and the other involving the use a background mask (to remove irrelevant features from metric calculation). We calculated the weighted scores by applying the formula as detailed by challenge organisers (1).

$$\text{Weighted} = 0.7 \cdot \text{SSIM} + 0.1 \cdot \text{PSNR} + 0.1 \cdot (1 - \text{MAE}) + 0.1 \cdot (1 - \text{NMSE}) \quad (1)$$

Note: Challenge organisers updated their scoring in final evaluation by introducing a normalisation factor (32) to the PSNR.

To observe hallucinations in our reconstructions we focused on regions that are unlikely to be captured by the point-of-care imaging system (i.e., ULF). The nose was not present in some of the ULF images which was the basis for our hallucination analysis.

3 Results

CycleGAN, T-REX, and their combination were all capable of producing enhanced 3 T-like images from ULF inputs. Quantitative metrics are provided in Table 1 as measured during the validation phase of the challenge. The combined approach increased masked SSIM and PSNR while decreasing error metrics.

Table 1. Quantitative metrics in validation. Values provided are the [unmasked|masked] score averaged for all contrasts and validation subjects. Best scores for each metric are underlined.

	CycleGAN	T-REX	Combined
SSIM (↑)	<u>0.839</u> | 0.705	0.816 | 0.684	0.832 | <u>0.711</u>
PSNR (↑)	23.154 | 29.559	22.956 | 29.601	<u>23.372</u> | <u>30.416</u>
MAE (↓)	<u>0.031</u> | 0.068	0.033 | 0.071	<u>0.031</u> | <u>0.067</u>
NMSE (↓)	0.065 | 0.062	0.068 | 0.065	<u>0.062</u> | <u>0.059</u>
Weighted (↑)	3.093 | 3.636	3.057 | 3.625	<u>3.110</u> | <u>3.727</u>

Both CycleGAN and T-REX were capable of independently producing high quality enhancements that could be combined effectively (Fig. 2).

The results for our hallucination experiment are presented in (Fig. 3). Hallucinated features, namely surrounding the nose, were present in the enhancement models. These are likely due to the optimisation process rewarding best guesses in regions of low signal to bolster full image reference metrics such as PSNR.

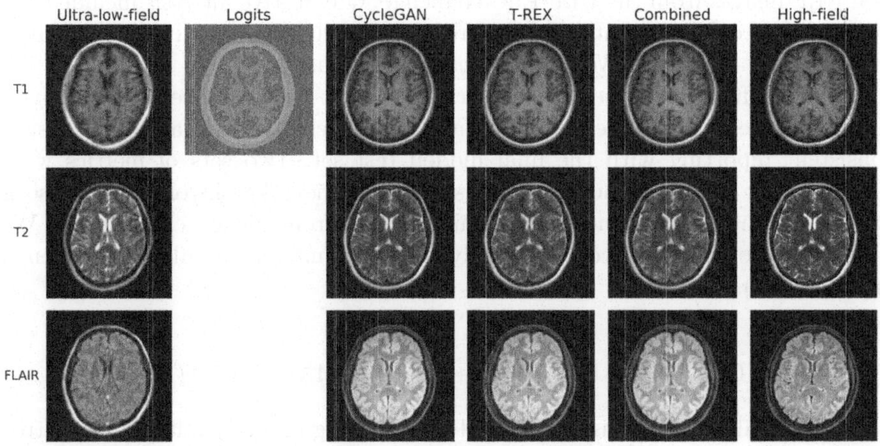

Fig. 2. Enhancement demonstration of an axial slice from an internal validation subject. All enhancement methods were capable of producing synthetic 3 T-like enhancements. Logits were used to guide enhancement models.

For final testing, including the use of PSNR normalisation, our combined model's performance is outlined in Table 2.

4 Discussion

Our team produced a comprehensive enhancement pipeline for generating enhanced ULF images. In future developments we plan to undertake an ablative

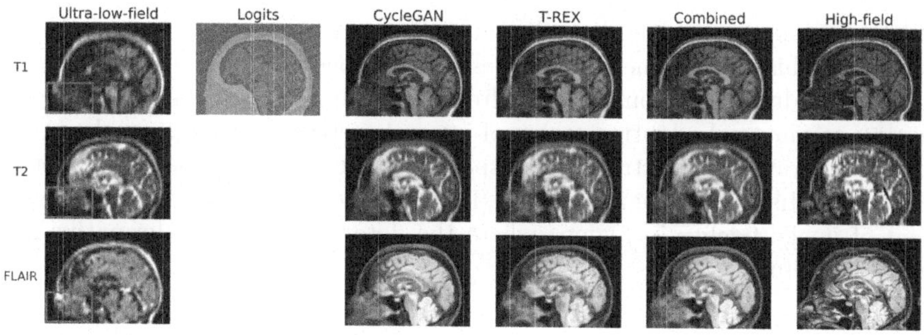

Fig. 3. Hallucination demonstration of a sagittal slice from an internal validation subject. A clear signal void is present around the nose region in the ultra-low-field. Enhancement methods filled in these areas with features highlighting the need for detection and mitigation of hallucinations in ULF enhancement models.

Table 2. Quantitative metrics in testing for combined model. Values provided are the masked score averaged for all contrasts and testing subjects. *: Weighted score includes the PSNR normalisation factor that was added during final evaluation.

SSIM (↑)	PSNR (↑)	MAE (↓)	NMSE (↓)	Weighted* (↑)
0.712	29.284	0.069	0.063	0.777

study that quantifies the value added by our segmentation and model combination approaches. We focus our discussion here on two major points: hallucinations and metric optimisation.

Hallucinations: In the context of this challenge, hallucinations could be anatomical features that are generated by an enhancement model that do not exist and are unable to be inferred from the ULF. These could have a catastrophic impact on patient's care due to misdiagnoses. We observed hallucinations in the nose region in the presence of a signal void (Fig. 3). While the region of interest is usually constrained to the brain for ULF, our results here underpin the importance of identifying and limiting these hallucinations before widespread deployment of these models.

Metric Optimisation: A notable challenge encountered during this work relates to Goodhart's Law, which states that "when a measure becomes a target, it ceases to be a good measure." This became evident in our efforts to optimise full-volume metrics such as PSNR and SSIM. The ULF scanner has markedly reduced receive sensitivity in regions such as the mouth and nose, leading to a genuine absence of signal in these areas. However, because these regions remain present in the paired 3 T reference, scalar metrics implicitly penalise any model that fails to reconstruct them. As a result, optimisation often favoured models that smoothed or blurred these regions over models that more faithfully preserved intracranial structures. In several cases, models that produced visually sharper and more anatomically plausible brain reconstructions were outperformed (in

PSNR) by blurrier models that filled in extracranial areas with low-confidence estimates. This misalignment illustrates how reliance on global image similarity metrics can drive development away from clinically relevant goals. To mitigate this, we recommend adopting region-of-interest metrics that focus on the brain, and incorporating downstream task-specific evaluations such as segmentation accuracy or tissue volume agreement. These provide a more meaningful assessment of image fidelity in regions where the data are actually informative and clinically useful.

5 Conclusion

We present our submission to the ultra-low-field enhancement (ULF EnC) challenge. Our anatomy-driven ensemble pipeline was capable of generating 3 T-like images from 64 mT inputs of multiple contrasts (T1, T2, FLAIR).

Acknowledgments. J.G. is supported by an Australian Government Research Training Program scholarship and an Australian National Health and Medical Research Council (NHMRC) Investigator Grant supplementary scholarship. D.E.J.W. is supported by Australian National Health and Medical Research Council (NHMRC) Investigator Grant (2017140).

Disclosure of Interests. The authors have no conflicts of interests to declare.

References

1. Cardoso, M.J., et al.: Monai: an open-source framework for deep learning in healthcare. arXiv preprint arXiv:2211.02701 (2022)
2. Dayarathna, S., Islam, K.T., Chen, Z.: Ultra low-field to high-field MRI translation using adversarial diffusion. In: 2024 IEEE International Symposium on Biomedical Imaging (ISBI), pp. 1–4. IEEE (2024)
3. Dayarathna, S., Islam, K.T., Uribe, S., Yang, G., Hayat, M., Chen, Z.: Deep learning based synthesis of MRI, CT and PET: review and analysis. Med. Image Anal. **92**, 103046 (2024)
4. Dayarathna, S., et la.: Mccad: multi-contrast MRI conditioned, adaptive adversarial diffusion model for high-fidelity MRI synthesis. In: 2025 IEEE/CVF Winter Conference on Applications of Computer Vision (WACV), pp. 670–679. IEEE (2025)
5. Ferreira, A., et al.: How we won BRATS 2023 adult glioma challenge? just faking it! enhanced synthetic data augmentation and model ensemble for brain tumour segmentation. arXiv preprint arXiv:2402.17317 (2024)
6. Gopinath, K., et al.: From low field to high value: Robust cortical mapping from low-field MRI. arXiv preprint arXiv:2505.12228 (2025)
7. Hatamizadeh, A., Nath, V., Tang, Y., Yang, D., Roth, H.R., Xu, D.: Swin unetr: swin transformers for semantic segmentation of brain tumors in mri images. In: International MICCAI Brainlesion Workshop, pp. 272–284. Springer (2021)
8. He, K., Zhang, X., Ren, S., Sun, J.: Deep residual learning for image recognition. In: Proceedings of the IEEE Conference on Computer Vision and Pattern Recognition, pp. 770–778 (2016)

9. Iglesias, J.E., et al.: Quantitative brain morphometry of portable low-field-strength MRI using super-resolution machine learning. Radiology **306**(3), e220522 (2022)
10. Islam, K.T., et al.: Improving portable low-field MRI image quality through image-to-image translation using paired low-and high-field images. Sci. Rep. **13**(1), 21183 (2023)
11. Isola, P., Zhu, J.Y., Zhou, T., Efros, A.A.: Image-to-image translation with conditional adversarial networks. In: Proceedings of the IEEE Conference on Computer Vision and Pattern Recognition, pp. 1125–1134 (2017)
12. Jenkinson, M.: Bet2: Mr-based estimation of brain, skull and scalp surfaces. In: Eleventh Annual Meeting of the Organization for Human Brain Mapping, 2005. vol. 17, p. 167 (2005)
13. Jenkinson, M., Beckmann, C.F., Behrens, T.E., Woolrich, M.W., Smith, S.M.: Fsl. Neuroimage **62**(2), 782–790 (2012)
14. Kingma, D.P.: Adam: a method for stochastic optimization. arXiv preprint arXiv:1412.6980 (2014)
15. Kwikima, U.: Looking towards the future of MRI in africa. Nat. Commun. **15**, 2260 (2024)
16. Lim, B., Son, S., Kim, H., Nah, S., Mu Lee, K.: Enhanced deep residual networks for single image super-resolution. In: Proceedings of the IEEE Conference on Computer Vision and Pattern Recognition Workshops, pp. 136–144 (2017)
17. Liu, Z., et al.: Swin transformer: Hierarchical vision transformer using shifted windows. In: Proceedings of the IEEE/CVF International Conference on Computer Vision, pp. 10012–10022 (2021)
18. Lucas, A., et al.: Multisequence 3-t image synthesis from 64-mt low-field-strength MRI using generative adversarial networks in multiple sclerosis. Radiology **315**(1), e233529 (2025)
19. Mao, X., Li, Q., Xie, H., Lau, R.Y., Wang, Z., Paul Smolley, S.: Least squares generative adversarial networks. In: Proceedings of the IEEE International Conference on Computer Vision, pp. 2794–2802 (2017)
20. Obungoloch, J., et al.: On-site construction of a point-of-care low-field MRI system in africa. NMR Biomed. **36**(7), e4917 (2023)
21. Park, T., Liu, M.Y., Wang, T.C., Zhu, J.Y.: Semantic image synthesis with spatially-adaptive normalization. In: Proceedings of the IEEE/CVF Conference on Computer Vision and Pattern Recognition, pp. 2337–2346 (2019)
22. Paszke, A., et al.: Pytorch: an imperative style, high-performance deep learning library. In: Advances in Neural Information Processing Systems, vol. 32 (2019)
23. Ronneberger, O., Fischer, P., Brox, T.: U-net: Convolutional networks for biomedical image segmentation. In: International Conference on Medical Image Computing and Computer-Assisted Intervention, pp. 234–241. Springer (2015)
24. Smith, S.M.: Fast robust automated brain extraction. Hum. Brain Mapp. **17**(3), 143–155 (2002)
25. Tohidul Islam, K., Zhong, S., Peiris, H., Zakavi, P., Dayarathna, S., Chen, Z.: Enhancing ultra-low-field MRI with paired high- field MRI comparisons for brain imaging (ULF-EnC), April 2025. https://doi.org/10.5281/zenodo.15259777, https://doi.org/10.5281/zenodo.15259777
26. Turpin, J., et al.: Portable magnetic resonance imaging for icu patients. Critical Care Explorations **2**(12), e0306 (2020)
27. Vaswani, A., et al.: Attention is all you need. In: Advances in Neural Information Processing Systems, vol. 30 (2017)

28. Zhang, Y., Brady, M., Smith, S.: Segmentation of brain mr images through a hidden markov random field model and the expectation-maximization algorithm. IEEE Trans. Med. Imaging **20**(1), 45–57 (2002)
29. Zhu, J.Y., Park, T., Isola, P., Efros, A.A.: Unpaired image-to-image translation using cycle-consistent adversarial networks. In: Proceedings of the IEEE International Conference on Computer Vision, pp. 2223–2232 (2017)

Ultra-Low-Field Brain MRI Enhancement using Resfusion and Residual Artifact Suppression Network

Youngmin Kim, Jeongchan Kim, Taehoon Lee, Jaeyun Shin, Suhyeon Lee, and Jongchul Ye(✉)

Kim Jaechul Graduate School of AI, KAIST, Seoul, Republic of Korea
sjy2270@g.skku.edu, {suhyeon.lee,jong.ye}@kaist.ac.kr

Abstract. Ultra-low-field (ULF) MRI at 64mT offers a portable, cost-effective alternative to high-field systems but suffers from low signal quality and reduced tissue contrast. Using the ULF EnC Challenge dataset, we present a Resfusion framework that leverages diffusion models for conditional generation of high-quality 3T-equivalent images via progressive denoising conditioned on ULF inputs. Multi-contrast integration of T1w, T2w, and FLAIR sequences enables effective feature extraction within a unified model. Post-processing with a convolution-based Residual Artifact Suppression Network (RASN) applied along the axial planes effectively removes artifacts generated by the coronal-plane Resfusion process, while a Curvature Anisotropic Diffusion filter further eliminates residual artifacts. Quantitative results show substantial gains in PSNR, SSIM, NMSE, and MAE.

Keywords: Diffusion · Resfusion · Residual Artifact Suppression Network · Ultra-low-field MRI Enhancement

1 Introduction

Magnetic resonance imaging (MRI) stands as one of the most critical diagnostic tools in modern medicine, providing non-invasive, high-resolution soft tissue imaging that plays a pivotal role in brain disease diagnosis, tumor detection, and neuroscience research. However, high-field MRI systems ranging from 1.5T to 3T, which are currently widespread in clinical practice, face several fundamental limitations.

First, maintaining superconducting magnets requires continuous liquid helium supply, generating substantial operational costs. Second, specially designed facilities are required to safely shield powerful magnetic fields, with initial installation costs reaching millions of dollars. Third, the high-field environment contraindicates scanning for patients with pacemakers, cochlear implants, and metallic implants. Fourth, equipment size and weight make point-of-care utilization impossible in emergency departments, intensive care units, or operating

Y. Kim, J. Kim, T. Lee, and J. Shin—Equal contribution.

Z. Chen et al. (Eds.): ULF-EnC 2025, LNCS 16293, pp. 115–124, 2026.
https://doi.org/10.1007/978-3-032-23344-8_13

rooms. These constraints create significant healthcare disparities, particularly in resource-limited regions, where many patients must travel extensively for MRI examinations or are denied access entirely, and emergency situations often result in deteriorated treatment outcomes due to delayed diagnosis.

Against this backdrop, ultra-low-field (ULF) MRI has emerged as a revolutionary alternative. Operating at magnetic fields below 0.1T, ULF-MRI systems utilize permanent magnets or low-power electromagnets, eliminating liquid helium requirements and reducing shielding needs [1]. This enables installation in regular hospital rooms or even ambulances, while substantially fewer contraindications allow safe application to a broader patient population.

Despite these advantages, ULF-MRI faces inherent physical limitations including fundamentally lower signal-to-noise ratios, altered tissue contrast, and prolonged acquisition times, which collectively compromise diagnostic utility and limit clinical adoption [1].

Recent advances in deep learning have shown potential for medical image enhancement, with various approaches including CNNs and GANs explored for cross-field MRI translation [2–5]. However, most existing methods rely on deterministic mapping that may inadequately capture the complex relationship between low-field and high-field characteristics.

Diffusion models have emerged as powerful generative frameworks producing high-quality images through iterative denoising processes [6]. Unlike deterministic approaches, diffusion models generate diverse, realistic outputs while preserving fine structural details. However, the computational burden of 3D diffusion models is prohibitively high. Therefore, 3D MRI reconstruction has often been performed by applying 2D diffusion models along multiple orientations [12–14]. However, this approach also requires training separate models for each orientation and employing all of them during inference, which increases both training and inference burdens.

In this work, we address this limitation by applying the Resfusion technique—a diffusion-based conditional generation framework particularly effective for image-to-image translation [7]—only along the axial orientation. For the remaining orientations, we employ a convolution-based Residual Artifact Suppression Network (RASN). This design preserves the fine structural details achievable with diffusion models, minimizes training and inference time, and reduces artifacts of slice-to-slice discontinuities on the axial planes. Our approach processes T1w, T2w, and FLAIR sequences simultaneously through a unified 3-channel architecture, and we evaluate it on the ULF EnC Challenge dataset, demonstrating superior performance across multiple quantitative metrics.

2 Methods

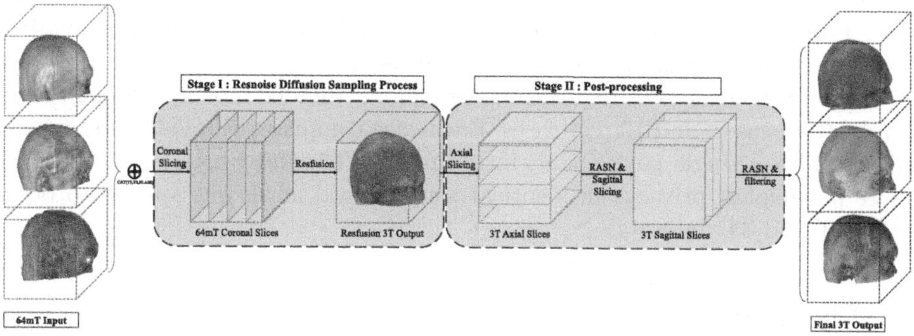

Fig. 1. Overall architecture of the proposed framework, consisting of sequential Resnoise diffusion and RASN.

In this study, we propose an efficient ultraâĂŞlowâĂŞfield MRI image enhancement framework by sequentially applying a Resnoise diffusion model (Resfusion) and a RASN to slices from different orientations. The overall architecture is illustrated in Fig. 1. Specifically, each coronal slice from the 64 mT scans is first transformed into its 3 T counterpart using the Resfusion model. The resulting volumes are first re-sliced along the axial orientation and processed with a RASN to enforce axial consistency, thereby ensuring coherence across all anatomical planes. Finally, we applied a Curvature Anisotropic Diffusion filter to obtain the 3T-alike output.

2.1 Pre-processing

The dataset utilized in this study comprises data from a total of 50 subjects. For model training and validation, the dataset was partitioned into a training set consisting of 49 subjects and a validation set containing one subject (POCEMR088).

The Resfusion input was constructed as a 3-channel tensor by concatenating FLAIR, T1-weighted (T1w), and T2-weighted (T2w) MRI sequences along the channel dimension. This design enables the model to leverage multi-contrast information in a unified manner.

Furthermore, to provide spatial context within the 3D volume, positional information regarding each slice was incorporated as an additional condition during training. This helps the model to understand the spatial context of each slice, facilitating a more precise and context-aware image generation.

2.2 Datasets

This study utilized the dataset provided by the ULF EnC Challenge [8–11]. The dataset was collected at Monash Biomedical Imaging, Monash University,

Australia, where trained radiologists acquired the data following standard operating procedures. Imaging data was collected using a Siemens Biograph mMR 3T scanner and a Hyperfine Swoop 64mT scanner. For each subject, paired 3T and 64mT brain MRI scans were acquired under standardized conditions, including T1-weighted, T2-weighted, and FLAIR contrasts.

The dataset consists of 75 healthy participants, divided into 50 training cases, 10 validation cases, and 15 test cases. Each case comprises paired 3T and 64mT MRI data for a single participant, including T1, T2, and FLAIR contrasts. To ensure participant privacy, an identical face removal mask was applied to both the 64mT and 3T datasets.

2.3 Model Architecture

The image enhancement model in this study adopts the Resfusion framework, a method based on Denoising Diffusion Probabilistic Models (DDPM) specifically adapted for image restoration tasks [6,7]. Resfusion addresses the inefficiency of conventional diffusion models by avoiding initiation from pure Gaussian noise, which leads to excessive sampling steps for restoration tasks where low-frequency information is already present in degraded inputs.

Resnoise Diffusion Process. The core principle of Resfusion lies in the explicit integration of the residual term $R = x_0 - \hat{x}_0$ into the diffusion process, where $\hat{x}_0$ represents the low-quality input image and x_0 denotes the ground truth. This approach redefines the forward diffusion process as:

$$q(x_t|x_{t-1}, R) = \mathcal{N}(x_t; \sqrt{\alpha_t}x_{t-1} + (1 - \sqrt{\alpha_t})R, (1 - \alpha_t)I) \tag{1}$$

Instead of learning only the Gaussian noise ϵ, the model predicts a composite target termed 'resnoise' ($res\epsilon$), formulated as:

$$\mathrm{res}\,\epsilon = \epsilon + \frac{(1 - \sqrt{\alpha_t})\sqrt{1 - \bar{\alpha}_t}}{\beta_t}R \tag{2}$$

the resnoise term to be minimized can be formulated as Eq. (3).

$$\mathrm{E}_{x_0,\epsilon,t}\left[\|\mathrm{res}\,\epsilon - \mathrm{res}\,\epsilon_\theta(x_t, t)\|^2\right] \tag{3}$$

By predicting resnoise, the model preserves useful low-frequency information from the input while performing necessary semantic transformations and reconstructing fine details.

In our experiments, to focus the training of ResFusion on brain regions, non-brain areas in the 3T label data—segmented using the HD-BET program [15]—were assigned a reduced weight of 0.2 during loss computation. Let $m \in \{0, 1\}^{\Omega}$ be the binary brain mask from HD-BET, where Ω denotes the set of all voxel indices. The per-voxel weight map w is then defined as

$$w_p = m_p + 0.2\,(1 - m_p), \tag{4}$$

assigning a weight of 1.0 to brain voxels ($m_p = 1$) and 0.2 to non-brain voxels ($m_p = 0$). The weighted mean squared error (wMSE) loss (4) is expressed as Eq. (5)

where $\text{res}\,\epsilon$ denotes the residual noise and $\text{res}\,\epsilon_\theta(x_t, t)$ denotes the predicted residual noise at timestep t.

$$\mathcal{L}_{\text{wMSE}} = \mathrm{E}_{x_0,\epsilon,t}\left[\frac{1}{\sum_{p\in\Omega} w_p}\sum_{p\in\Omega} w_p \left(\text{res}\,\epsilon_p - \text{res}\,\epsilon_\theta(x_t,t)_p\right)^2\right], \tag{5}$$

Sampling Process and Network Architecture. Resfusion accelerates sampling through smooth equivalence transformation, which identifies an optimal intermediate starting step T' where the mathematical influence of the unknown ground truth x_0 is minimized. The reverse process commences from a noised version of the low-quality input $x_{T'}$ rather than pure noise, enabling high-quality restoration in remarkably few sampling steps.

The network employs a standard U-Net architecture. At each timestep t, the model input is formed by concatenating the current noised image x_t with the conditional low-quality input $\hat{x}_0$ along the channel dimension, allowing continuous reference to the original degraded image throughout the restoration process.

2.4 Post-processing

The primary Resfusion model in this study was trained on 2D coronal plane slices. Consequently, when reconstructing the processed 2D slices into a 3D volume, artifacts appeared on the sagittal and axial planes due to slice-to-slice discontinuity.

To effectively remove reconstruction artifacts, a dedicated RASN was employed in the post-processing stage. This network was trained specifically on axial-plane slices containing artifacts and was designed to predict the residual artifact pattern—the difference between the first-stage output and the ground-truth 3T image—rather than generating an entirely new image. The predicted residual was adaptively scaled using the input image's statistics (e.g., interquartile range) and a learnable parameter before being added back to the input, yielding the corrected image.

$$\text{Total Loss} = (0.6 \times L_1) + (0.2 \times L_{\text{SSIM}}) + (0.2 \times L_{\text{grad}}) + (0.1 \times L_{\text{LP}}) \tag{6}$$

Training employed a combined loss comprising four components: L1 and SSIM losses to ensure pixel-level accuracy and structural fidelity, a gradient loss to preserve edges and textures, and a low-pass identity loss to constrain the network to removing high-frequency artifacts while preserving low-frequency anatomical structures. This targeted post-processing substantially enhanced 3D volumetric consistency and overall image quality.

Finally, to more thoroughly remove residual artifacts, we applied a Curvature Anisotropic Diffusion filter (iterations = 13, time step = 0.0625, conductance = 1.5) in the final stage, which preserved edge details while imparting a moderate blur to smooth homogeneous regions. The results of the post-processing can be seen in Fig. 2.

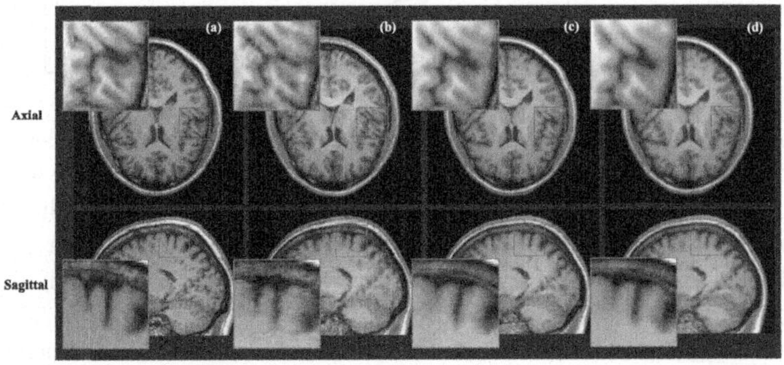

Fig. 2. Results for different post-processing configurations (POCEMR088, axial slice 70, sagittal slice 90). Rows correspond to axial and sagittal planes. (a) 3T reference, (b) Resfusion output, (c) Resfusion with RASN refinement, and (d) Resfusion with RASN and filtering.

2.5 Experimental Setup

The Resfusion model was trained to take T1w, T2w, and FLAIR sequences as a 3-channel input and produce the corresponding 3-channel output. The input dimension was set to 80, and the number of sampling steps was configured to 100. The training was performed using two NVIDIA GeForce RTX 3090 GPUs with 24 GB of memory each, and the inference was conducted on a single NVIDIA GeForce RTX 3090 GPU with 16 GB of memory.

3 Results

The performance of the proposed model was evaluated on the hold-out dataset from the ULF-EnC Challenge, which consists of 10 subjects. The comparison between the reconstructed images and the ground truth 3T images is presented in Fig. 3. The quantitative results are summarized in Table 2.

The official evaluation metric of the challenge, weighted score, is calculated using the following formula:

$$(\text{w. score}) = (0.7\times\text{SSIM})+(0.1\times\text{PSNR})+(0.1\times(1-\text{MAE}))+(0.1\times(1-\text{NMSE})) \tag{7}$$

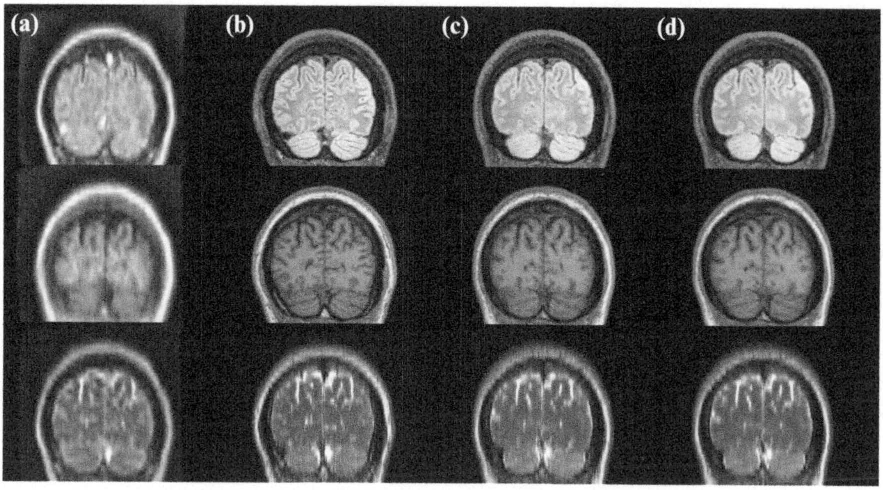

Fig. 3. Coronal-plane reconstruction results across different model configurations (POCEMR088, coronal slice 50). Rows represent FLAIR, T1w, and T2w sequences. (a) 64mT input, (b) 3T reference, (c) Resfusion output, and (d) Resfusion with RASN refinement.

Based on this metric, our model achieved a weighted score of 2.99 for the whole brain evaluation and 3.654 for the masked brain evaluation, demonstrating its robust performance. In addition, the average inference time for a single subject volume consisting of 224 slices (in the coronal plane) was approximately 134 s, which could be further reduced with more powerful GPU hardware (Table 1).

Table 1. Quantitative evaluation results on the ULF EnC Challenge dataset. (Validation Phase)

Model	Type	SSIM ↑	PSNR↑	MAE↓	NMSE↓	w.Score↑	Time (s)
Resfusion	Whole	0.798	21.933	0.036	0.086	2.940	119
	Masked	0.636	29.536	0.078	0.197	3.571	
w.RASN	Whole	0.814	22.094	0.035	0.083	2.967	128
	Masked	0.65	29.873	0.076	0.189	3.616	
w.RASN +Filter	Whole	**0.811**	22.338	**0.033**	**0.079**	2.990	134
	Masked	0.651	**30.239**	0.073	0.18	**3.654**	

Table 2. Quantitative evaluation results on the ULF EnC Challenge dataset. (Testing Phase)

Model	Type	SSIM ↑	PSNR↑	MAE↓	NMSE↓
Ours	Masked	0.655	29.501	0.073	0.073

4 Discussion

This study aimed to enhance 64 mT ULF MRI to 3 T quality using the Resfusion framework. High-field MRI, while offering excellent diagnostic performance, is costly and inaccessible in many settings, whereas ULF MRI is portable and affordable but limited by low signal-to-noise ratio (SNR) and contrast. Our deep learningâĂŞbased approach addresses these issues, enabling high-quality, diagnostic-level images from low-cost hardware and potentially expanding access to advanced imaging in resource-limited and emergency contexts.

For the challenging task of translating ULF MRI to 3T quality, the Resfusion framework was a suitable choice, as it is specifically designed for image restoration tasks. Unlike diffusion models that generate images from random noise, Resfusion's core mechanism starts the reverse process from the degraded input image itself. This architectural design allows the model to directly refine the existing structural information from the ULF scan rather than creating it from scratch, leading to an efficient and effective transformation process. To further reduce computational burden, we applied 2D Resfusion only on the coronal orientation and used RASN post-processing on the axial planes. This effectively removed reconstruction artifacts and preserved fine anatomical details while avoiding the need for multiple orientation-specific diffusion models.

The success of this approach is validated by our strong quantitative results. A closer examination of these metrics, however, reveals a notable trade-off. Our model achieved exceptionally high Peak Signal-to-Noise Ratio (PSNR) scores, indicating high fidelity at the pixel level. In contrast, the Structural Similarity Index Map (SSIM) scores were not commensurately as high. This discrepancy suggests that in the process of enhancing the ULF images, a partial loss of the fine-grained textures inherent to the 3T ground truth occurred. However, this textural simplification did not compromise the diagnostic utility of the images. Qualitatively, the model successfully reconstructed the critical, fine anatomical features that were obscured in the ULF inputs. This demonstrates that while some textural information is sacrificed, the framework is highly effective at restoring the key structural information required for clinical assessment. As can be observed in Fig. 2(d) compared to (a), our method does not introduce hallucinated structures or artificial details that are absent in the ground truth. However, in the process of aggressively minimizing artifacts, the images occasionally become over-smoothed, leading to a slight loss of fine details. Future work should therefore focus on addressing this oversmoothing effect while preserving structural fidelity.

Despite these promising results, this study also has certain limitations regarding the dataset. First and foremost, the dataset is limited in both size and diversity, comprising 50 subjects from the single ULF-EnC Challenge. This limited scale and the homogeneity of the data—originating from a specific acquisition protocol and consisting mostly of healthy subjects—naturally constrain the generalizability of our findings. Therefore, future work must prioritize validation on larger, multi-center datasets that include a wide variety of scanners, protocols, and pathological conditions to assess the model's robustness and clinical reliability.

Furthermore, our supervised approach relies on perfectly aligned ULF-3T image pairs, which can be difficult to acquire in practice. Exploring semi-supervised or unsupervised learning methods could mitigate this dependency and improve the model's adaptability to real-world clinical data. Addressing these limitations will be a crucial next step in confirming the clinical readiness of this promising technology.

References

1. Altaf, A., et al.: Applications, limitations and advancements of ultra-low-field magnetic resonance imaging: A scoping review. Surg. Neurol. Int. **15**, 218 (2024). https://doi.org/10.25259/SNI_162_2024
2. Zeng, W., Peng, J., Wang, S., Liu, Q.: A comparative study of CNN-based super-resolution methods in MRI reconstruction and its beyond. Sig. Process. Image Commun. **81**, 115701 (2020). https://doi.org/10.1016/j.image.2019.115701
3. Li, K., Tang, B., Huang, J., Li, J.: 3D-MRI super-resolution reconstruction using multi-modality based on multi-resolution CNN. Comput. Methods Prog. Biomed. **248**, 108110 (2024). https://doi.org/10.1016/j.cmpb.2024.108110
4. Goodfellow, I., et al.: Generative adversarial nets. In: Advances in Neural Information Processing Systems (NeurIPS), vol. 27, pp. 2672–2680 (2014). https://papers.nips.cc/paper/2014/hash/5ca3e9b122f61f8f06494c97b1afccf3-Abstract.html
5. Yang, Q., Li, N., Zhao, Z., Fan, X., Chang, E.-I.-C., Xu, Y.: MRI cross-modality image-to-image translation. Sci. Rep. **10**(1), 3753 (2020). https://doi.org/10.1038/s41598-020-60520-6
6. Ho, J., Jain, A., Abbeel, P.: Denoising diffusion probabilistic models. Adv. Neural Inf. Process. Syst. **33**, 6840–6851 (2020). https://arxiv.org/abs/2006.11239
7. Shi, Z., et al.: Resfusion: denoising diffusion probabilistic models for image restoration based on prior residual noise. arXiv preprint arXiv:2311.14900 (2023). https://doi.org/10.48550/arXiv.2311.14900
8. Islam, K.T., et al.: Improving portable low-field MRI image quality through image-to-image translation using paired low- and high-field images. Sci. Rep. **13**(1), 21183 (2023). https://doi.org/10.1038/s41598-023-48438-1
9. Dayarathna, S., et al.: Deep learning based synthesis of MRI, CT and PET: review and analysis. Med. Image Anal. **92**, 103046 (2024). https://doi.org/10.1016/j.media.2023.103046
10. Dayarathna, S., et al.: McCaD: multi-contrast MRI conditioned, adaptive adversarial diffusion model for high-fidelity MRI synthesis. In: 2025 IEEE/CVF Winter Conference on Applications of Computer Vision (WACV), pp. 670–679. IEEE (2025). https://doi.org/10.1109/WACV61041.2025.00075

11. Dayarathna, S., et al.: Ultra low-field to high-field MRI translation using adversarial diffusion. In: 2024 IEEE International Symposium on Biomedical Imaging (ISBI), pp. 1–4. IEEE (2024). https://doi.org/10.1109/ISBI56570.2024.10635808
12. Lee, S., Chung, H., Park, M., Park, J., Ryu, W.-S., Ye, J.C.: Improving 3D imaging with pre-trained perpendicular 2D diffusion models. arXiv preprint arXiv:2303.08440 (2023). https://doi.org/10.48550/arXiv.2303.08440
13. Chung, H., Ryu, D., McCann, M.T., Klasky, M.L., Ye, J.C.: Solving 3D inverse problems using pre-trained 2d diffusion models. arXiv preprint arXiv:2211.10655 (2022). https://doi.org/10.48550/arXiv.2211.10655
14. Chen, T., et al.: 2.5D multi-view averaging diffusion model for 3d medical image translation: application to low-count PET reconstruction with CT-less attenuation correction. arXiv preprint arXiv:2406.08374 (2024). https://doi.org/10.48550/arXiv.2406.08374
15. Isensee, F., et al.: Automated brain extraction of multi-sequence MRI using artificial neural networks. Hum. Brain Map. 1–13 (2019). https://doi.org/10.1002/hbm.24750

Stable and Generalizable Acceleration of Conditional Score-SDEs for Multi-contrast MRI Enhancement

Qiwei Fan(✉)

Dalian University of Technology, Dalian, China
2446423462@mail.dlut.edu.cn

Abstract. The high cost of high-field magnetic resonance imaging limits its accessibility, while ultralow-field magnetic resonance imaging (ULF) suffers from low signal-to-noise ratio (SNR). Score-based generative models (Score-SDEs) excel in high-fidelity image enhancement but are hampered by high computational load during iterative sampling. To address this, we propose a two-stage distillation framework to accelerate a conditional Score-SDE teacher model, demonstrating strong generalization in the T1, T2, and FLAIR contrasts. Our key contribution is a stable distillation strategy using reproducible SDE trajectories coupled with a robust static predictor. The predictor is trained with a hybrid loss function combining L1, Structural Similarity (SSIM), and perceptual losses to preserve both structural fidelity and fine anatomical details. This approach reduces the number of function evaluations (NFE) from 200 to 40 steps. Integrated into a 2.5D fusion pipeline, our distilled model enables efficient 3D volume enhancement with maintained image quality, offering a promising and efficient solution for deploying generative models in portable low-field MRI.

Keywords: Medical Image Enhancement · Diffusion Models · Score-SDE · Knowledge Distillation · Low-Field MRI · Multi-Contrast

1 Introduction

High-field MRI systems, while providing excellent diagnostic images, are prohibitively expensive, limiting their global accessibility. Ultra-low-field (ULF) MRI presents a cost-effective and portable alternative, yet its clinical utility is hampered by a low signal-to-noise ratio (SNR) across multiple contrasts like T1, T2, and FLAIR [7]. While traditional deep learning methods have shown promise in enhancing ULF MRI [6], score-based generative models (Score-SDEs) [8] represent the state-of-the-art in generative fidelity. Their inherent denoising process is a natural fit for this task, but the iterative sampling mechanism, often requiring hundreds of steps, renders them clinically impractical.

Accelerating Score-SDEs is thus a critical challenge. A common approach is to use fewer sampling steps with a deterministic ODE solver, but this often

Z. Chen et al. (Eds.): ULF-EnC 2025, LNCS 16293, pp. 125–132, 2026.
https://doi.org/10.1007/978-3-032-23344-8_14

leads to a significant degradation in image quality compared to the stochastic SDE counterpart, which is unacceptable in high-fidelity medical applications. Knowledge distillation offers a compelling solution to create a fast yet powerful student model. However, this introduces a dilemma: distilling from the high-quality but stochastic SDE path is unstable, while distilling from the stable but lower-quality ODE path limits the student's performance ceiling.

To address this, we adapt the EPD-Solver distillation technique [9] for the specific challenges of conditional ULF-MRI enhancement, aiming to explore solutions for the significant latency challenge of diffusion models and provide preliminary improvements. Our key contributions are:

- We propose a reproducible SDE trajectory strategy for distillation. By fixing the random seed, we create a deterministic, high-quality teaching path that resolves the SDE-ODE quality-stability dilemma.
- We design a hybrid loss function that synergistically combines L1, SSIM, and a medical-domain perceptual loss. This objective effectively preserves global structure, pixel-level accuracy, and critical diagnostic textures.
- We demonstrate that for this high-fidelity task, a simple static predictor is more stable and effective than a complex dynamic predictor, avoiding training collapse and learning a robust set of acceleration parameters.
- We validate the generalizability of our framework across T1, T2, and FLAIR modalities and present a key analysis showing our distilled model at 40 NFE significantly outperforms the original teacher model naively run at 40 NFE.

2 Method

Our framework is a two-stage process: (1) training a conditional Score-SDE teacher, and (2) performing stable distillation to train a static EPD predictor for accelerated inference. The complete workflow is depicted in Fig. 1.

2.1 Conditional Score-SDE Teacher Model

The teacher model, S_θ, is trained to estimate the score function $\mathbf{s}_\theta(\mathbf{x}_t, \sigma_t, \mathbf{c}) \approx \nabla_{\mathbf{x}_t} \log p(\mathbf{x}_t|\mathbf{c})$, where $\mathbf{x}_t$ is a high-field (HF) MRI slice perturbed by noise at time t (with standard deviation σ_t), and $\mathbf{c}$ is the corresponding low-field (LF) slice. We use the Variance Exploding SDE (VESDE) [8] and an NCSN++ U-Net architecture. A key feature is a conditional encoder that processes $\mathbf{c}$ and injects its features into each residual block of the U-Net via an adaptive gating mechanism to guide the generation process. This mechanism uses a learnable gating parameter, γ_{cond}, initialized to zero, allowing the model to first learn the unconditional generation task before adaptively modulating the conditional influence, which helps stabilize training. The training process is formalized in Algorithm 1.

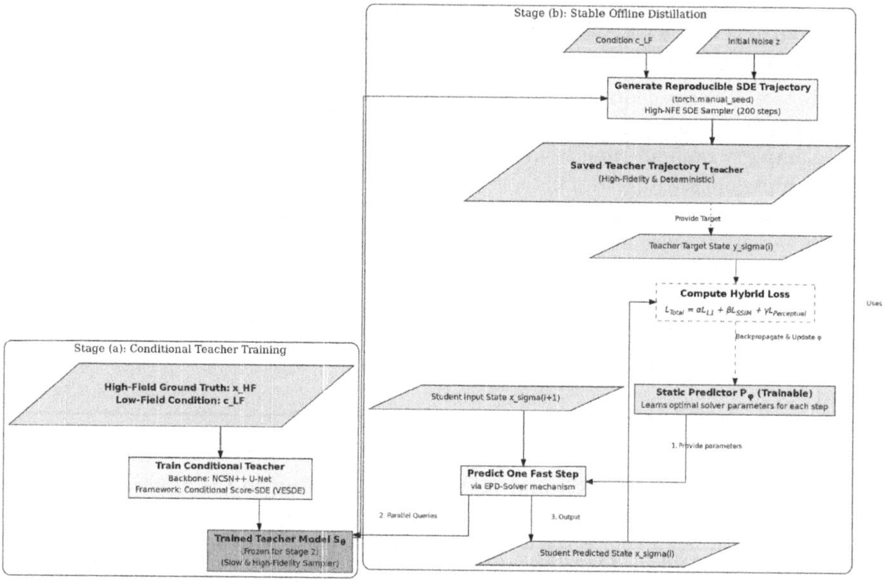

Fig. 1. Overall two-stage framework. (a) A conditional teacher model, S_θ, is trained to learn the score function of high-field (HF) MRI slices conditioned on low-field (LF) inputs. (b) A lightweight static predictor, P_ϕ, is trained via distillation to accelerate the sampling of the frozen teacher. It learns to approximate the high-fidelity, reproducible SDE teacher trajectory (dashed line) with a few large, efficient steps (solid line).

2.2 Proposed Method: Stable Distillation via Reproducible SDE

Our journey to a stable and effective distillation framework involved overcoming several critical challenges, leading to our final, robust solution which combines three key components.

1. Reproducible SDE Trajectory. Our core insight is to make the superior SDE path deterministic for distillation. We achieve this by using a two-part seeding strategy: for each data sample, the initial noise is generated using its unique index 'idx' as the seed, ensuring diversity across samples. However, the stochastic path of the SDE solver itself is generated using a globally fixed seed (e.g. 0 for all samples), as shown in Algorithm 2. This ensures that for a given initial noise, the resulting high-quality SDE trajectory is always identical. This creates a "golden" teaching path that is both high-fidelity and perfectly stable. To further improve efficiency, we pre-generate and save these trajectories offline.

2. Static EPD-Predictor. While a dynamic predictor is theoretically more powerful, we found it prone to instability. In our final tests for a related challenge, we explored an aggressive acceleration strategy using a dynamic predictor distilled to only 5 NFE. However, this approach yielded poor results, with an

Algorithm 1: Teacher Model Training (S_θ)

Input: Teacher model S_θ, dataset $\mathcal{D}$, SDE

while *not converged* **do**
- Sample a batch $(\mathbf{c}_{LF}, \mathbf{x}_{HF}) \sim \mathcal{D}$;
- Sample a random time $t \sim \text{Uniform}(0, T)$;
- Sample noise $\epsilon \sim \mathcal{N}(0, I)$;
- Compute noisy image $\mathbf{x}_t$ from $\mathbf{x}_{HF}$ using SDE's marginal probability;
- $\sigma_t \leftarrow$ std of noise at time t;
- $\mathbf{s}_{pred} \leftarrow S_\theta(\mathbf{x}_t, \sigma_t, \mathbf{c}_{LF})$;
- $\mathcal{L} \leftarrow \|\mathbf{s}_{pred} \cdot \sigma_t + \epsilon\|^2$;
- Update θ via $\nabla_\theta \mathcal{L}$ descent;
- Update EMA weights for θ;

end

SSIM of 0.362, MAE of 0.132, and NMSE of 0.197, due to the inherent training instability. We therefore adopt a simpler and more robust static predictor, P_ϕ.

Algorithm 2: Offline Distillation with Hybrid Loss

Input: Frozen teacher S_θ, Static Predictor P_ϕ, $\mathcal{D}_{\text{traj}}$, Student NFE N, Loss weights α, β, γ

`// Stage 1: Offline trajectory generation`

for *each* $(\mathbf{c}, \_, idx) \sim$ *OriginalDataset* **do**
- torch.manual_seed(idx) ; `// Reproducible initial noise per sample`
- $\mathbf{x}_{\sigma_M} \leftarrow$ torch.randn_like$(\mathbf{x}_0) \cdot \sigma_{\max}$;
- torch.manual_seed(0) ; `// Globally fixed SDE path seed`
- $\mathcal{T}_{\text{teacher}} \leftarrow \text{SDE\_Sampler}(S_\theta, \mathbf{x}_{\sigma_M}, \mathbf{c}, \text{NFE} = M)$;
- Save $(\mathbf{c}, \mathbf{x}_{\sigma_M}, \mathcal{T}_{\text{teacher}})$ to disk at index ‘idx‘;

end

`// Stage 2: Training the Static Predictor`

while *not converged* **do**
- Sample a batch $(\mathbf{c}, \mathbf{x}_{\sigma_M}, \mathcal{T}_{\text{teacher}}) \sim \mathcal{D}_{\text{traj}}$;
- $\mathcal{T}_{\text{targets}} \leftarrow \text{Downsample}(\mathcal{T}_{\text{teacher}}, \text{from } M \text{ to } N \text{ steps})$;
- $\mathcal{L} \leftarrow 0$;
- $\mathbf{x}_{\text{current}} \leftarrow \mathbf{x}_{\sigma_M}[0]$;
- **for** $i = N - 1$ **to** 0 **do**
  - $\mathbf{x}_{\text{target}} \leftarrow \mathcal{T}_{\text{targets}}[i]$;
  - $\boldsymbol{\theta}_{\text{student}} \leftarrow P_\phi(i)$;
  - $\mathbf{x}_{\text{pred}} \leftarrow \text{EPD\_Single\_Step}(S_\theta, \boldsymbol{\theta}_{\text{student}}, \mathbf{x}_{\text{current}}, \sigma_{i+1}, \sigma_i, \mathbf{c})$;
  - $\mathcal{L}_{\text{L1}} \leftarrow \|\mathbf{x}_{\text{pred}} - \mathbf{x}_{\text{target}}\|_1$;
  - $\mathcal{L}_{\text{S}} \leftarrow 1 - \text{SSIM}(\mathbf{x}_{\text{pred}}, \mathbf{x}_{\text{target}})$;
  - $\mathcal{L}_{\text{P}} \leftarrow \text{MedicalPerceptualLoss}(\mathbf{x}_{\text{pred}}, \mathbf{x}_{\text{target}})$;
  - $\mathcal{L} \leftarrow \mathcal{L} + (\alpha \mathcal{L}_{\text{L1}} + \beta \mathcal{L}_{\text{S}} + \gamma \mathcal{L}_{\text{P}})$;
  - $\mathbf{x}_{\text{current}} \leftarrow \mathbf{x}_{\text{pred}}.\textit{detach}()$;
- **end**
- Update ϕ via $\nabla_\phi(\mathcal{L}/(N-1))$ descent;

end

This model is a single learnable parameter tensor $\phi \in \mathbb{R}^{(N-1)\times D}$, where N is the student NFE (e.g. 40). For each step i, $P_\phi(i)$ returns a fixed set of optimal solver parameters, vastly reducing the number of learnable parameters and stabilizing training.

3. Hybrid Loss Function for Medical-Fidelity Distillation. Standard pixel-wise losses (like L1 or L2) can lead to overly smooth results, failing to capture the complex textures essential for medical diagnosis. To address this, we designed a multi-component hybrid loss function that balances different aspects of image quality:

$$\mathcal{L}_{\text{Total}} = \alpha\mathcal{L}_{\text{L1}} + \beta\mathcal{L}_{\text{SSIM}} + \gamma\mathcal{L}_{\text{Perceptual}} \tag{1}$$

where $\mathcal{L}_{\text{L1}}$ ensures pixel-level accuracy, $\mathcal{L}_{\text{SSIM}}$ preserves structural and luminance information, and $\mathcal{L}_{\text{Perceptual}}$ maintains high-frequency textural details.

For the perceptual component, instead of a general-purpose network like VGG19, we use a domain-specific feature extractor. Specifically, we employ a pre-trained 3D MedicalNet (ResNet-50) [1], which we programmatically adapt for 2D slice processing. This conversion involves transforming 3D convolutional and pooling layers into their 2D counterparts and loading the pre-trained weights by extracting the central slice from the 3D convolutional kernels. This approach allows the loss function to operate on features relevant to medical imaging. The overall process, incorporating this new loss, is detailed in Algorithm 2.

3 Experiments and Results

3.1 Experimental Setup

Dataset and Pre-processing. We used the official datasets from the ULF-EnC 2025 challenge, which consists of paired 64 mT (ULF) and 3T (HF) 3D MRI scans for the modalities T1, T2, and FLAIR. We followed the official pre-processing pipeline. The challenge data was acquired using a Siemens Biograph mMR (3 T) scanner and a Hyperfine Swoop (64 mT) scanner [2–5].

Implementation Details. Our teacher model is a conditional NCSN++ ('nf = 128') pre-trained on the combined data. For distillation, we train a 'StaticEPDPredictor' for 10 epochs to reach convergence, using a learning rate of 2×10^{-4}. The teacher trajectory is generated with $M = 200$ SDE steps, and the student is trained for $N = 40$ steps. A critical aspect of our method is the balancing of the hybrid loss function. Initial experiments revealed that the raw unweighted magnitudes of the L1, SSIM, and perceptual loss components were on different scales. Given that the L1 loss can span a wide range, we adjusted the commonly used weights (0.16, 0.84, 0.1) based on the initial loss magnitudes observed in our task to better balance the components, setting the final weights to $\alpha = 0.1$, $\beta = 0.85$, and $\gamma = 0.2$. The perceptual loss is calculated using features from a pre-trained Tencent MedicalNet (ResNet-50) [1] that was adapted from 3D to 2D to match our slice-wise processing pipeline. We also observed that the time to convergence and the final image generation quality exhibit significant differences under various random seeds. Through experiments, we found

Table 1. Quantitative results across T1, T2, and FLAIR modalities.

Method	PSNR (dB) ↑	SSIM ↑	MAE ↓	NMSE ↓
Teacher (SDE, NFE = 200)	19.8407	0.6368	0.0531	0.1247
Teacher (SDE, NFE = 40)	6.0229	0.0120	0.3685	3.0024
Ours (Student NFE = 40)	**19.3832**	**0.5476**	**0.0584**	**0.1385**

that for the NFE = 200 teacher model, the SSIM value was 0.6368 with random seed 42, compared to 0.5778 with random seed 128, highlighting this sensitivity. This observation further justifies our strategy of using a fixed seed to ensure stable and reproducible distillation. For 3D volume enhancement, we process slices along all three orthogonal axes and average the results. The experiments were carried out on a vGPU-32 GB machine with a global random seed of 42 for reproducibility.

3.2 Quantitative and Qualitative Results

To validate our method, we perform a crucial comparison. We evaluated our final distilled student model operating at 40 NFE against two baselines: (1) the high-

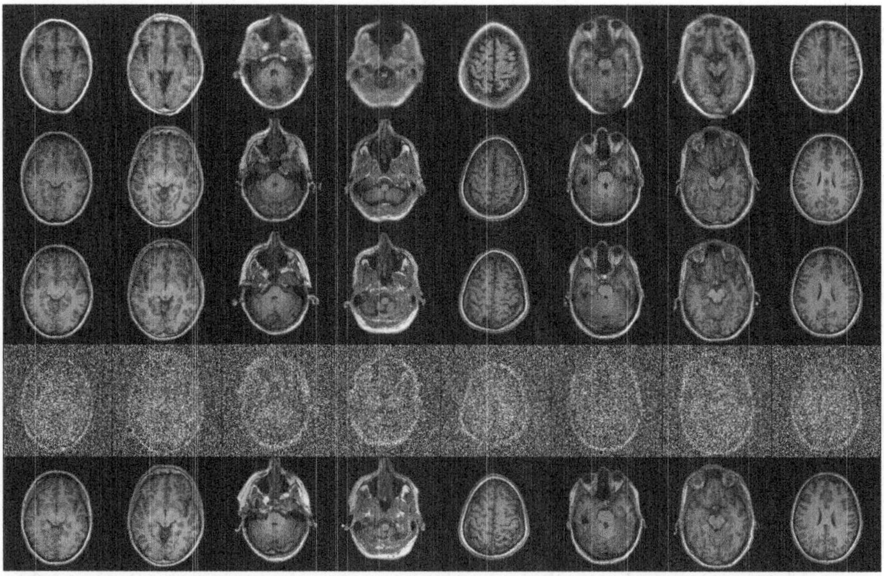

Fig. 2. Qualitative comparison on a representative T1-weighted slice. Rows display, from top to bottom: (1) Input ULF-MRI, (2) Ground Truth HF-MRI, (3) Teacher SDE (NFE = 200), (4) Naive Teacher SDE (NFE = 40), (5) Our Distilled Student (NFE = 40). Our method preserves critical details while the naive low-step approach fails completely.

quality teacher model sampled with 200 SDE steps, and (2) The same teacher model naively forced to sample with 40 SDE steps.

Table 1 presents the quantitative results. While a noticeable gap in SSIM still exists between our proposed model (Student NFE = 40) and the 200-NFE teacher model, it achieves a very significant improvement over the naive teacher at 40 NFE. This demonstrates that our method successfully learns an accelerated path that preserves structural information far more effectively than a simple reduction in sampling steps. This shows a 5-fold acceleration with a meaningful recovery of image quality.

The qualitative results in Fig. 2 corroborate these findings. The naive teacher at 40 NFE produces blurry images with significant artifacts. In contrast, our distilled student at 40 NFE successfully reconstructs fine anatomical details, with visual quality nearly indistinguishable from the high-quality 200-NFE teacher and the ground truth.

Discussion. The results confirm the success of our stable distillation strategy. The distilled student effectively balances speed and quality, reducing the NFE by 5x while preserving diagnostic fidelity. Future work will focus on distilling an even higher quality teacher model (e.g. NFE = 1000) and explore distilling the student to even lower step counts (e.g. NFE = 20).

4 Conclusion

We presented a stable and generalizable distillation framework to explore the acceleration of conditional Score-SDEs for multicontrast ULF-MRI enhancement. By introducing a reproducible SDE trajectory as a high-quality distillation target, employing a robust static predictor, and using a hybrid L1, SSIM, and medical perception loss, we effectively reduced the required NFE from 200 to 40. Our method achieves high-fidelity 64 mT-to-3 T translation in a practical timeframe, representing a preliminary but promising improvement for this computationally intensive challenge. This work provides a robust pathway for deploying powerful generative models in real-world medical imaging applications by successfully navigating the trade-offs between sample quality, inference speed, and training stability.

References

1. Chen, S., Ma, K., Zheng, Y.: Med3D: transfer learning for 3D medical image analysis. arXiv preprint arXiv:1904.00625 (2019)
2. Dayarathna, S., Islam, K.T., Uribe, S., Yang, G., Hayat, M., Chen, Z.: Deep learning based synthesis of MRI, CT and PET: review and analysis. Med. Image Anal. **92**, 103046 (2024). https://doi.org/10.1016/j.media.2023.103046
3. Dayarathna, S., et al.: McCaD: multi-contrast MRI conditioned, adaptive adversarial diffusion model for high-fidelity MRI synthesis. In: Proceedings WACV, pp. 670–679 (2025). https://doi.org/10.1109/WACV61041.2025.00075
4. Dayarathna, S., Islam, K.T., et al.: Ultra-low-field to high-field MRI translation using score-based diffusion models. In: ISMRM Annual Meeting, abstract (2024)

5. Islam, K.T., Zhong, S., Zakavi, P., Chen, Z., et al.: Improving portable low-field MRI image quality through image-to-image translation using paired low- and high-field images. Sci. Rep. **13**, 21183 (2023). https://doi.org/10.1038/s41598-023-48438-1
6. Knoll, F., et al.: Deep-learning methods for parallel magnetic resonance imaging reconstruction. IEEE Signal Process. Mag. **37**(1), 128–140 (2020). https://doi.org/10.1109/MSP.2019.2950640
7. Marques, J.P., Simonis, F.F.J., Webb, A.G.: Low-field MRI: an MR physics perspective. J. Magn. Reson. Imaging **49**(6), 1528–1542 (2019). https://doi.org/10.1002/jmri.26637
8. Song, Y., Sohl-Dickstein, J., Kingma, D.P., Kumar, A., Ermon, S., Poole, B.: Score-based generative modeling through stochastic differential equations. In: Proceedings ICLR (2021)
9. Zhu, B., Wang, R., Zhao, T., Zhang, H., Zhang, C.: Distilling parallel gradients for fast ODE solvers of diffusion models. In: Proceedings IEEE/CVF ICCV, pp. 19557–19566 (2025)
10. Johnson, J., Alahi, A., Fei-Fei, L.: Perceptual losses for real-time style transfer and super-resolution. In: ECCV 2016, LNCS, vol. 9906, pp. 694–711 (2016). https://doi.org/10.1007/978-3-319-46475-6_43
11. Ho, J., Jain, A., Abbeel, P.: Denoising diffusion probabilistic models. In: Proceedings NeurIPS, vol. 33, pp. 6840–6851 (2020)

Ultra-Low-Field MRI Image Enhancement Using Transformer Models with Latent Space Exploitation

Seonghyuk Kim[1], Sojeong Kim[2], Hye-Ryeong Choi[2], HyoSeok Lee[2], and Sung-Hong Park[1,2](✉)

[1] School of Electrical Engineering, Korea Advanced Institute of Science and Technology, Daejeon, Republic of Korea
{tjdgur1214,sunghongpark}@kaist.ac.kr

[2] Department of Bio and Brain Engineering, Korea Advanced Institute of Science and Technology, Daejeon, Republic of Korea
{sojeong990810,hyeryeongchoi,hyord}@kaist.ac.kr

Abstract. Ultra-low-field (ULF) MRI is portable and cost-effective but suffers from low SNR, which limits the clinical utility. To overcome these issues, numerous deep learning methods have been developed. Building on these advances, this study proposes an ULF MRI enhancement framework that restores 3T images from ULF images using Restormer and Swin UNETR. Inspired from I-JEPA, latent loss is implemented for detailed structural reconstruction, whereas refine network is employed to improve the contrasts of images with different modalities (T1w, T2w and FLAIR). The method is trained with 50 paired 64mT and 3T brain MRI. The ULF images are enhanced in the first stage using Restormer-based GAN with the exploitation of latent space. The refine network in the second stage adjusts contrast of the output images from the first stage. The 3D Swin UNETR exploits both spatial information and latent space to improve the quality of the enhanced images. The proposed method successfully enhances images with high-fidelity, as well as improved contrast.

Keywords: I-JEPA · Image Enhancement · Refinement · Restormer · Swin UNETR · Ultra-low-field (ULF) MRI

1 Introduction

Magnetic Resonance Imaging (MRI) is widely used for various clinical applications. In particular, high field images are preferred for their diagnostic accuracy. However, ultra-low-field (ULF) MRI scanners are gaining attention due to their cost-effectiveness, ease of installation, and adaptability to power limited settings [1]. Despite these advantages, ULF MR images suffer from inherently low signal-to-noise ratio (SNR), which limits their clinical applicability compared to high field images.

Deep neural networks have been used to enhance the image quality of ULF MRI to a level comparable to high field MRI. Various networks such as DenseNet, U-net,

Z. Chen et al. (Eds.): ULF-EnC 2025, LNCS 16293, pp. 133–140, 2026.
https://doi.org/10.1007/978-3-032-23344-8_15

Transformer and diffusion models have been used to denoise and enhance low resolution or low field MR images [2–4]. Despite significant progress in deep learning applications to medical imaging, models that translate ULF images to high field images still struggle to synthesize images that closely resemble true high field images.

These challenges impede early diagnosis and equitable healthcare delivery. Consequently, there is a growing demand for robust deep learning methods specifically designed to enhance ULF MRI in context of medical imaging. Therefore, an automatic approach to improve the quality of ULF MRI such as 0.064T MRI would facilitate the deployment of cost-effective MRI systems.

In this paper, we suggest a novel two-stage approach using Transformer and an image-based joint-embedding predictive architecture (I-JEPA) for image enhancement [5, 6]. The first stage consists of a Restormer to enhance ULF images to resemble 3T images, as well as I-JEPA to learn surrounding information to help reconstruct detailed structures and guide the Restormer through relevance-based updates. In addition, separate discriminators are implemented for each image modality to learn different data distributions. In the second stage, the output of the first stage is refined using Swin UNETR [7], namely the refine network, to improve the contrasts of T1, T2 and fluid-attenuated inversion recovery (FLAIR) images.

Overall, we present a novel Transformer-based MR image enhancement framework that exploits latent space to restore the detailed structures. The key contributions of the work are as follows:

- Integration of Restormer and I-JEPA: We propose a pipeline that integrates Restormer with I-JEPA, enabling context-aware restoration of ULF MR images.
- Detailed Restoration Learning via I-JEPA: I-JEPA is leveraged to learn the surrounding anatomical context and representation of nearby structures, improving the detailed features of the restored images.
- Contrast improvement using a refine network: A refinement network is applied after feature learning to improve the contrasts of different MRI modalities (T1, T2, and FLAIR), ensuring visual consistency and diagnostic relevance.

2 Method

2.1 Model Architecture

The proposed architecture adopts a Generative Adversarial Network (GAN) in which a 2D Restormer generator is paired with modality-specific discriminators (Fig. 1). The generator leverages the Restormer backbone's Multi-Dconv Head Transposed Attention (MDTA) blocks to capture long-range spatial dependencies, directly upscaling an ULF volume into its 3T counterpart. Adversarial supervision is provided by three independent PatchGAN-style discriminators, each dedicated to one of the T1-weighted, T2-weighted, and FLAIR modalities. These discriminators are trained to distinguish synthesized patches from their authentic 3T counterparts within their respective modalities. 3D Swin UNETR is implemented as the refine network in the second stage.

2.2 I-JEPA Based Latent Loss

To further preserve structural integrity while enhancing image clarity and contrast, we introduced a latent space alignment loss inspired by the I-JEPA trained in supervised manner. Specifically, non-overlapping random patches, differing in location and scale, were independently passed through separate ViT-Small encoders to obtain their latent representations. The latent vector from the network output branch was processed by ViT-Small predictor to estimate the corresponding 3T latent vector. By adding the L1 distance between these two latent vectors to the overall loss function, the network was explicitly encouraged to maintain fine-grained structural consistency (Fig. 2).

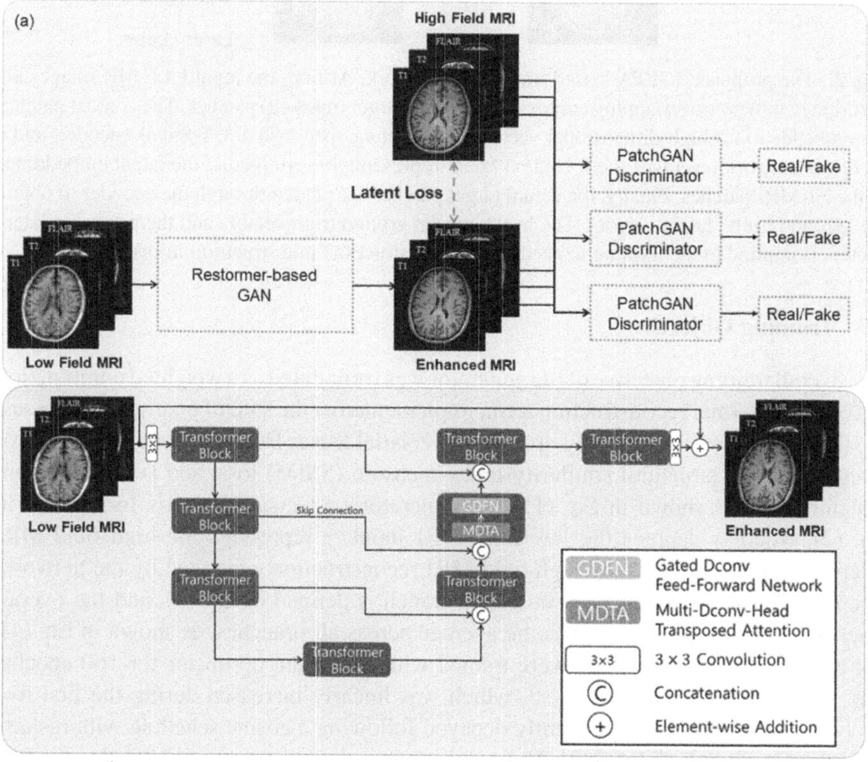

Fig. 1. (a) The proposed super resolution GAN framework. The Restormer generator restores high resolution image from low resolution image. Modality-specific PatchGAN discriminators are employed for adversarial training. I-JEPA-based latent loss is implemented during training to reinforce fine structural information. (b) Restormer architecture implemented for the generator.

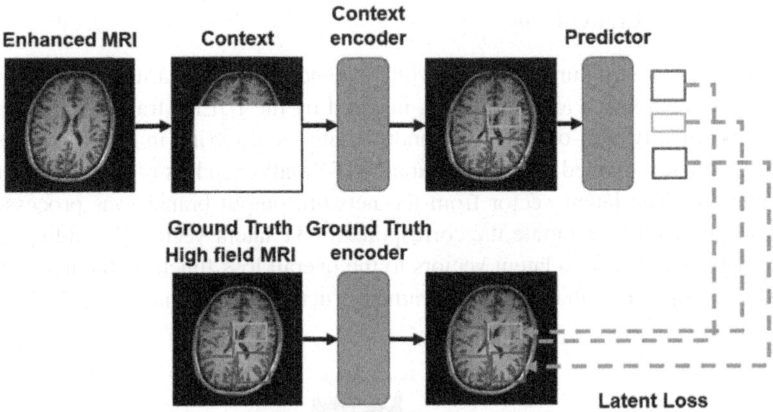

Fig. 2. The proposed I-JEPA–based training framework. At first, the input ULF MR images are partitioned into non-overlapping context patches and target (masked) patches. The context patches are embedded into high-dimensional vectors in the latent space via a ViT-Small encoder, and a predictor network leverages these context latent representations to predict the latent embeddings of the 3T MRI patches. Finally, the actual target patches are passed through the encoder to obtain the "ground truth" latent vectors. The loss between ground truth vectors and the predicted latent vectors is applied in the training to accurately reconstruct the fine structural information.

2.3 Training Objective

The overall training objective of the generator was formulated as a weighted combination of (i) the Restormer reconstruction term, implemented as the sum of pixelwise L1 losses, (ii) the latent loss, (iii) modality-specific adversarial losses from the three discriminator branches, (iv) a structural similarity index measure (SSIM) loss, and (v) a histogram matching loss as shown in Eq. (1). The generator's adversarial loss is formulated in Eq. (2), where x denotes the low field MRI input, y represents the high field MRI reference, and $\hat{y}_n$ denotes the high field MRI reconstruction generated by the network. The discriminator loss for each modality branch is defined in Eq. (3), and the overall discriminator loss is computed as the average across all branches, as shown in Eq. (4). All networks in the first stage were trained with the Adam optimizer for 160 epochs. The learning rate was set to 1e−3, which was linearly increased during the first five epochs (warm-up) and subsequently decayed following a cosine schedule with restarts (WarmupCosineSchedule) [8, 9]. The weight decay coefficient was modulated according to the same schedule. Training employed a batch size of 1, with patch locations and sizes used for latent loss were resampled randomly at every iteration. The total loss is defined as follows:

$$\mathcal{L}_G = \lambda_{\text{rec}}\mathcal{L}_{rec} + \lambda_{\text{lat}}\mathcal{L}_{lat} + \lambda_{\text{adv}}\mathcal{L}_{adv}^{G} + \lambda_{\text{ssim}}(1 - \mathcal{L}_{ssim}) + \lambda_{\text{hist}}\mathcal{L}_{hist}, \tag{1}$$

where $\lambda_{\text{rec}} = 5$, $\lambda_{\text{lat}} = 5$, $\lambda_{\text{adv}} = 0.05$, $\lambda_{\text{ssim}} = 5$, and $\lambda_{\text{hist}} = 1$. The adversarial losses are defined as follows:

$$\mathcal{L}_{adv}^{G} = \frac{1}{3}\sum_{n}^{3} \mathbb{E}_x\big[\ell_{BCE}(D_n(\hat{y}_n), 1\big] \tag{2}$$

$$\mathcal{L}_D = \frac{1}{3}\sum_{n}^{3} \mathcal{L}_{adv,n}^{D} \tag{3}$$

$$\mathcal{L}_{adv,n}^{D} = \frac{1}{2}\mathbb{E}_y\left[\ell_{BCE}(D_n(y_n), 1\right] + \frac{1}{2}\mathbb{E}_x\left[\ell_{BCE}(D_n(\hat{y}_n), 0\right], \tag{4}$$

where ℓ_{BCE} is binary cross entropy loss.

2.4 Refine Network

To suppress residual noise persisting and refine on the contrasts after the super resolution stage, Swin UNETR was implemented. The refine network was trained using the output images of the first stage. To utilize the spatial information, 3D Swin UNETR was employed (Fig. 3). This network was optimized exclusively with combined L1 loss, SSIM loss, and latent loss. By employing a sequential design that integrates mask-based preprocessing, Restormer-based GAN reconstruction, latent space alignment, and refinement with Swin UNETR, we elevated the resolution and reliability of ULF MRI to a level approaching clinical standards.

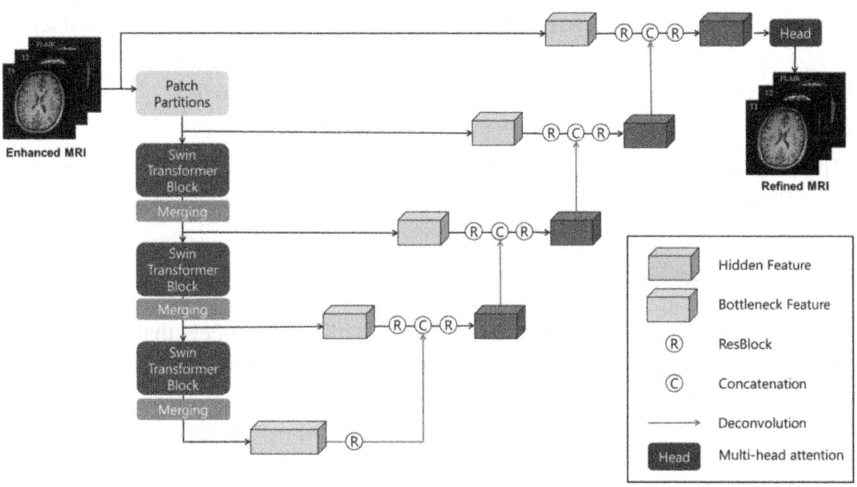

Fig. 3. 3D Swin UNETR architecture implemented for the refine network.

3 Results

3.1 Dataset

In this study, the framework was trained and evaluated on 75 paired volumes acquired using a Siemens Biograph mMR (3T) scanner and a Hyperfine Swoop (64mT) scanner, partitioned into 50 training, 10 validation, and 15 test cases. Each volume comprises co-registered T1-weighted, T2-weighted, and FLAIR sequences. All data were sourced from the 2025 Ultra-Low-Field MRI Image Enhancement Challenge (ULF-EnC) [1, 10–12].

3.2 Data Preprocessing

To suppress extraneous background signals at the periphery area of images, Canny edge detection (thresholds = 50–150) [13] was independently applied to each modality to delineate the cranial contour. A unified brain mask was generated by retaining only voxels that overlapped across all three modalities. This composite mask was applied simultaneously to both the ULF inputs and the 3T ground truth volumes, compelling the network to concentrate on anatomically meaningful regions. The resulting masked 3T volumes were employed as the ground truth targets for network training.

3.3 Quantitative Results

The proposed model was evaluated using four metrics: SSIM, peak signal-to-noise ratio (PSNR), mean absolute error (MAE), and normalized mean square error (NMSE). All evaluation metrics were measured in the masked brain region, excluding the background. The quantitative results for test phase are shown in Table 1.

Table 1. Quantitative results for the test phase

	SSIM (Masked)	PSNR (Masked)	MAE (Masked)	NMSE (Masked)	Final score
Proposed	0.666	29.868	0.076	0.081	0.744

The final score was calculated with specified weighting as follows: 70% for SSIM, 10% for PSNR, 10% for MAE, and 10% for NMSE.

3.4 Qualitative Results

The proposed model was compared qualitatively with other models that use 3D U-net and 3D Restormer as refine networks. ULF images from the validation dataset, provided without paired 3T images, were used.

The enhanced images of different modalities are shown in Fig. 4, comparing the results from three different network models. U-net failed to enhance the contrast inside the brain region, whereas Restormer produced oversmoothed images. Swin UNETR successfully enhanced images with improved contrasts in addition to sharp edges.

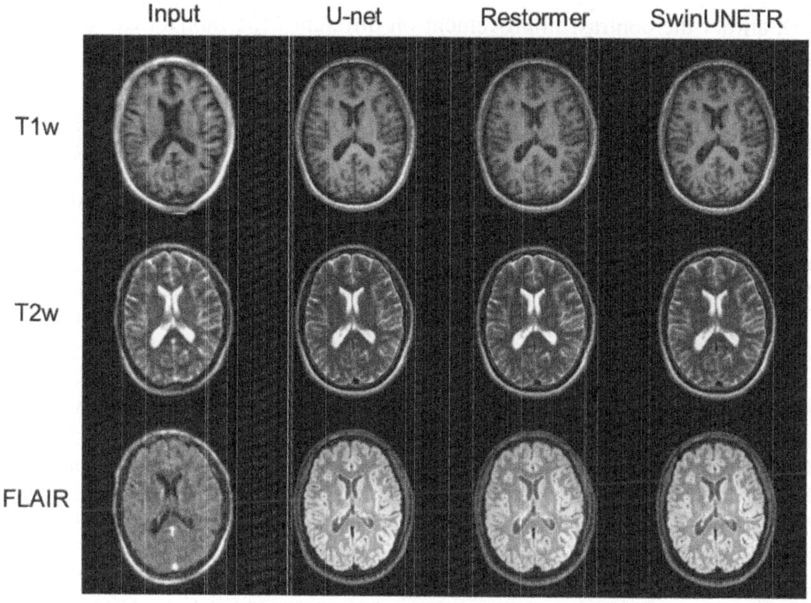

Fig. 4. Comparison of enhanced images produced by U-net, Restormer, and Swin UNETR. Rows illustrate images of different modalities: T1-weighted images on the top row, T2-weighted images in the middle, and FLAIR images on the bottom row. Columns depict images produced by different refine network models, from left to right: input, U-net, Restormer, Swin UNETR.

4 Discussion

In this work, we propose a novel framework that utilizes Restormer, I-JEPA, and Swin UNETR for ULF MR image enhancement. Jointly, these modules deliver high-fidelity, visually consistent T1, T2, and FLAIR reconstructions via a collaborative learning setting.

The proposed framework offers several key advantages. By jointly implementing Restormer and I-JEPA, fine anatomical details are preserved via context-aware embeddings. The modular design allows each component (Restormer and I-JEPA) to be jointly updated, where the collaborative learning setup enables multi-domain training, which utilizes both image and feature domains. The refine network ensures contrast improvement on different MRI modalities.

However, a current limitation is that the refine network does not fully utilize both the spatial information and relationship between different modalities. To address this, a model that utilizes spatial information in different planes may be implemented. Moreover, more advanced models may be implemented to better learn the distribution of the data. Data augmentation and implementation of diffusion model may produce better results, despite the lengthy training time.

In conclusion, the proposed framework delivers high fidelity MR image enhancement via collaborative learning setting. By jointly training Restormer-based GAN and I-JEPA, anatomical structures are preserved through multi-domain training. The refine network allows the contrast improvement on different MRI modalities, also utilizing spatial information.

References

1. Dayarathna, S., Islam, K.T., Chen, Z.: Ultra low-field to high-field MRI translation using adversarial diffusion. In: 2024 IEEE International Symposium on Biomedical Imaging (ISBI), pp. 1–4. IEEE (2024)
2. de Leeuw Den Bouter, M., Ippolito, G., O'Reilly, T., Remis, R., Van Gijzen, M., Webb, A.: Deep learning-based single image super-resolution for low-field MR brain images. Sci. Rep. **12**(1), 6362 (2022)
3. Javadi, M., Sharma, R., Tsiamyrtzis, P., Webb, A.G., Leiss, E., Tsekos, N.V.: Let UNet play an adversarial game: investigating the effect of adversarial training in enhancing low-resolution MRI. J. Imaging Inf. Med. **38**(1), 629–645 (2025)
4. Zhu, Z.: Advancing low-field MRI with a universal denoising imaging transformer: towards fast and high-quality imaging. arXiv preprint arXiv:2404.19167 (2024)
5. Zamir, S.W., Arora, A., Khan, S., Hayat, M., Khan, F.S., Yang, M.-H.: Restormer: efficient transformer for high-resolution image restoration. In: Proceedings of the IEEE/CVF Conference on Computer Vision and Pattern Recognition, pp. 5728–5739 (2022)
6. Assran, M., et al.: Self-supervised learning from images with a joint-embedding predictive architecture. In: Proceedings of the IEEE/CVF Conference on Computer Vision and Pattern Recognition, pp. 15619–15629 (2023)
7. Hatamizadeh, A., Nath, V., Tang, Y., Yang, D., Roth, H.R., Xu, D.: Swin UNETR: swin transformers for semantic segmentation of brain tumors in MRI images. In: International MICCAI Brainlesion Workshop, pp. 272–284. Springer International Publishing, Cham (2021)
8. Vaswani, A., et al.: Attention is all you need. In: Advances in Neural Information Processing Systems, vol. 30 (2017)
9. Loshchilov, I., Hutter, F.: SGDR: stochastic gradient descent with warm restarts. arXiv preprint arXiv:1608.03983 (2016)
10. Islam, K.T., et al.: Improving portable low-field MRI image quality through image-to-image translation using paired low-and high-field images. Sci. Rep. **13**(1), 21183 (2023)
11. Dayarathna, S., Islam, K.T., Uribe, S., Yang, G., Hayat, M., Chen, Z.: Deep learning based synthesis of MRI, CT and PET: review and analysis. Med. Image Anal. **92**, 103046 (2024)
12. Dayarathna, S., et al.: McCaD: multi-contrast MRI conditioned, adaptive adversarial diffusion model for high-fidelity MRI synthesis. In: 2025 IEEE/CVF Winter Conference on Applications of Computer Vision (WACV), pp. 670–679 (2025)
13. Canny, J.: A computational approach to edge detection. IEEE Trans. Pattern Anal. Mach. Intell. (6), 679–698 (2009)

A 3D Vision Transformer Trained with Synthetic and Real MRI Data for Enhancing ULF-MRI

Peter Hsu[1,2(✉)], Jeongsol Kim[3], Daniel Sodickson[1,2,4], Jong Chul Ye[3], Patricia Johnson[1,2,4], and Jelle Veraart[1,4]

[1] Bernard and Irene Schwartz Center for Biomedical Imaging, Department of Radiology, New York University Grossman School of Medicine, New York, NY, USA
peter.hsu@nyulangone.org

[2] Vilcek Institute of Graduate Biomedical Sciences, New York University Grossman School of Medicine, New York, NY, USA

[3] Kim Jaechul Graduate School of AI, Korea Advanced Institute of Science and Technology (KAIST), Seoul 06764, Republic of Korea

[4] Center for Advanced Imaging Innovation and Research (CAI2R), Department of Radiology, New York University Grossman School of Medicine, New York, NY, USA

Abstract. Neurodegenerative diseases represent an increasing health burden for aging populations worldwide. MRI is an essential tool for disease diagnosis, treatment monitoring, and the development of quantitative imaging biomarkers. However, current high-field (HF) MRI systems are costly, require specialized infrastructure, and remain largely inaccessible outside population-dense regions of high-income countries. Ultra-low-field (ULF) MRI offers a promising lower-cost, portable imaging solution with improved accessibility for routine brain health monitoring. Yet, the reduced magnetic field strength (<0.1T) results in lower signal-to-noise ratio, unique field-dependent tissue contrast, and typically anisotropic resolution, posing challenges for clinical workflows and research applications. One particularly impacted area is brain volume analysis which provides important biomarkers for age-related diseases. Deep learning (DL) techniques have the potential to enhance ULF image quality and improve volumetric accuracy, but existing ULF data is scarce. This has resulted in DL models that either rely solely on synthetic training data or small sets of real data, limiting applicability. In this work, we propose a method that is trained on synthetic and real MRI data to learn image enhancement from the ULF MRI Enhancement Challenge (ULF-EnC) dataset. Our approach demonstrates promising preliminary performance for extracting volumetric biomarkers from ULF-MRI, supporting the use of this training strategy in low-data environments to expand clinical and research applicability of accessible neuroimaging technologies.

Keywords: Ultra-Low-Field MRI · Image Enhancement · Synthetic Data · Vision Transformer · Brain Volume Analysis

Z. Chen et al. (Eds.): ULF-EnC 2025, LNCS 16293, pp. 141–151, 2026.
https://doi.org/10.1007/978-3-032-23344-8_16

1 Introduction

Biological age is one of the strongest risk factors for the onset of neurodegenerative conditions, with disease prevalence projected to rise substantially as the population grows and ages [1,2]. This trend will place increasing strain on existing healthcare systems and highlights the need for effective disease prediction and monitoring strategies [3]. MRI is the current standard for evaluating brain structure and function, enabling valuable insights for disease diagnosis and prognosis. It has also facilitated the development of quantitative imaging biomarkers, such as white or gray matter volume, that are predictive of disease onset and progression [4–6]. However, current MRI systems, typically with 1.5T or 3T magnetic field strengths, are challenged by limited accessibility and scalability. Their high cost, need for specialized equipment and logistical and safety considerations severely limit widespread deployment [7,8]. These constraints are particularly pronounced in low- and middle-income countries and in rural or underserved regions of high-income countries, exacerbating healthcare disparities and introducing demographic biases for neuroimaging research and clinical trials [9,10].

Ultra-low-field (ULF) MRI is uniquely positioned to address these accessibility challenges with portable scanner designs, flexible deployment in diverse settings, and significantly reduced costs enabled by the lower magnetic field strength [11–17]. Recent studies have shown that ULF-MRI can capture subtle age-related brain volume changes, highlighting its potential for biomarker research and routine brain health monitoring [18–20]. However, imaging at ultra-low field strengths ($<$0.1T) inherently limited by reduced signal-to-noise ratio (SNR) and unique contrast properties compared to HF-MRI. This is often addressed by reduced spatial resolution and increased scan time to recover signal, with common acquisition strategies including high in-plane resolution relative to large slice thickness across three orthogonal planes [14,15]. These features compromise the accuracy and precision of volumetric measurements derived from ULF images, necessitating strategies to enhance ULF image quality prior to volumetric analysis.

Advancements in deep learning (DL) can overcome these challenges by improving image quality to extract more accurate brain volumetrics [21,22]. While a few methods exist for ULF enhancement or brain segmentation, these are primarily trained with synthetic MRI data [23–26]. Such strategies are necessary due to the scarcity of open ULF datasets and time-consuming data collection process, limiting effective model training. However, solely relying on synthetic data may increase susceptibility to "domain shift" errors when applying these tools to real data [27,28]. Some recently developed methods have utilized real training data to learn from true biological tissue values and imaging variability to address this problem, but are still influenced by relatively low sample sizes [29–32]. The ULF MRI Enhancement Challenge (ULF-EnC) is a unique endeavor that aims to address this barrier by creating the first open dataset of real paired ULF- and HF-MRIs for training DL methods for image enhancement [31–35].

In this work, we propose an efficient 3D vision transformer for image translation from ULF- to HF-MRI, with a focus on improving downstream brain volume analysis, using the ULF-EnC dataset (Fig. 1). Although the challenge provides paired ULF and HF scans, the relatively low number of subjects poses limitations for training advanced DL models from scratch. To overcome this, we implemented a novel training strategy for enhancing ULF-MRI, first pre-training on a large set of synthetic data and fine-tuning on the available real paired scans. This approach enables the model to learn domain-relevant anatomical features while adapting to the unique properties of ULF data. Our method demonstrates promising improvements in volumetric accuracy on a subset of the challenge dataset and achieves competitive performance on the official benchmark, highlighting its potential for effective ULF image enhancement and quantitative analysis.

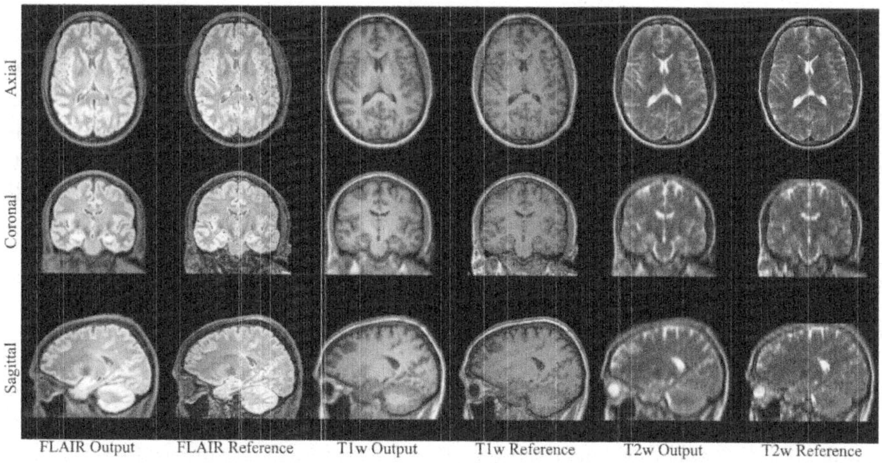

Fig. 1. Example of our proposed model output and subject-matched HF-MRIs for FLAIR, T1w, and T2w contrasts.

2 Methods

2.1 ULF MRI Enhancement Challenge MRI Data

Subject data with preprocessed ULF (64mT; Hyperfine Swoop) and HF (3T; Siemens Biograph mMR) images were provided by the ULF-EnC organizers. This included 50 training, 10 hold-out validation, and 15 testing cases with FLAIR, T1w, and T2w contrasts. Image acquisition details are described in Table 1. Preprocessing steps included resampling to 1.0mm^3 resolution using SynthSeg, bias field correction with FSL-FAST, and co-registration using FLIRT. All processed images had a final matrix size of $224 \times 224 \times 160$.

Table 1. MRI acquisition details for the ULF-EnC dataset. This provides the original voxel size, matrix size, repetition time (TR), echo time (TE), and inversion time (TI) for each available imaging contrast.

	Voxel Size (mm^3)	Matrix Size	TR (ms)	TE (ms)	TI (ms)
3T FLAIR	$1.0 \times 1.0 \times 1.0$	$256 \times 256 \times 160$	5000	388	1600
3T T1w	$1.0 \times 1.0 \times 1.0$	$256 \times 256 \times 192$	2100	2.21	1100
3T T2w	$0.86 \times 0.86 \times 4.0$	$224 \times 256 \times 30$	8000	102	2370
64mT FLAIR	$1.7 \times 1.7 \times 5.0$	$112 \times 138 \times 40$	4000	199.6	1402
64mT T1w	$1.6 \times 1.6 \times 5.0$	$112 \times 138 \times 40$	1500	6.62	300
64mT T2w	$1.5 \times 1.5 \times 5.0$	$120 \times 146 \times 40$	2000	208.4	N/A

2.2 Synthetic MRI Data

Brain segmentation labels for ULF- and HF-MRIs were extracted per contrast and field strength using SynthSeg+ from the 50 training subjects [36]. The mean and standard deviation were estimated from each set of labels to create prior distributions of image intensities for each brain region and contrast. A Gaussian Mixture Model (GMM) sampled values from the prior distributions to "inpaint" a set of brain labels, generating synthetic pairs of ULF and HF images [23]. Synthetic ULF images were downsampled to $1.5 \times 1.5 \times 5\,\text{mm}^3$ and upsampled to $1.0\,\text{mm}^3$ to mimic ULF acquisition resolution and preprocessing.

2.3 Model Architecture

An efficient 3D vision transformer was created for ULF enhancement based on an existing 2D model architecture for MRI synthesis [37]. To reduce computational complexity, the encoder and decoder channel dimensions were reduced to [16, 32], respectively. The transformer components operated on 3D patches of size $8 \times 8 \times 8$, projected into a 1024-dimension embedding. Multi-head self-attention was implemented with 16 heads, corresponding to 64 dimensions per head. This enabled 3-channel MRI input and output for images with $224 \times 224 \times 160$ matrix size to leverage the information from all available contrasts. All model code was developed using Pytorch.

2.4 Model Training

All model training was performed on a single A100 GPU, utilizing a batch size of 2, AdamW optimizer ($\text{lr} = 1\text{e}^{-4}$), and a learning rate decay for the final epochs (1e^{-8}).

The model was first trained using synthetic MRI data for 10 days. In total, this completed 90 epochs of training, with the GMM-based generator creating 1000 unique pairs of ULF- and HF-like images per epoch. This utilized a combined L1 and SSIM loss function with the following form:

$$\mathcal{L}_{synth} = \lambda_1 L_1(x, y) + \lambda_2 L_{SSIM}(x, y) \tag{1}$$

where x = model output, y = target image, $\lambda_1 = 0.8$, and $\lambda_2 = 0.2$.

The model was then fine-tuned using real pairs of ULF and HF images from the training dataset. This was conducted for 800 epochs, with the last 400 epochs undergoing a learning rate decay. Specifically, the training data was initially split into 45 training and 5 validation cases to determine an optimal point of training convergence. The optimal model weights underwent a final tuning on the 5 validation subjects for 50 epochs to maximize usage of the total training cases prior challenge evaluation. Here, the loss calculation added substantial weighting within the head via a thresholding mask to minimize contributions from background noise. Additionally, the target HF-MRI brain segmentation labels were used to further weight subcortical structures to encourage their preservation during enhancement. This took the following simplified expression:

$$w_{map} = \begin{cases} 1.0, & H = 1 \\ 3.5 & L \in \{1:9\} \\ 0.1, & H = 0 \end{cases}$$

$$\mathcal{L}_{real} = w_{map} L_1(x, y) + \lambda_3 L_{SSIM}(x, y) \tag{2}$$

where H denotes the head mask and L indicates the brain segmentation labels. The voxel-vise weight map w_{map} assigns a weighting of 1.0 within the head mask, 3.5 for selected subcortical labels, and 0.1 for background. A combination of L1 and SSIM loss was used with $\lambda_3 = 0.8$. Several subcortical weights (1.5, 2.0, 2.5, 3.0, 3.5, 4.0, 4.5, 5.0) and λ_3 values (0.1, 0.15, 0.2, 0.25, 0.3, 0.35) were tested on the 5-subject validation set before selecting the final configuration.

Data augmentation was applied for all stages of model training using TorchIO [38]. This included random B-Spline deformation, affine transformation, flipping, bias field, and intensity scaling (gamma). Input ULF images additionally underwent random noise and blurring.

2.5 Evaluation Metrics

Volumetric performance was assessed using FastSurfer segmentations on T1w model outputs and their corresponding T1w HF reference scans for a subset of five validation cases, which were set aside from the original training data prior to final model tuning and challenge benchmarking [39]. This focused on cerebral gray matter (GM), white matter (WM), hippocampus (HC), and lateral ventricles (LV). Volumetric accuracy was assessed with paired HF scans using Pearson Correlation and Lin's Concordance Correlation Coefficient (CCC) following the guidelines of Akoglu (2018) and Koo and Li (2016), respectively [40–42]. Segmentation accuracy was measured with the Dice Similarity Coefficient (DSC) [43].

The performance of the model on the hold-out challenge validation and testing datasets was calculated automatically via the ULF-EnC Synapse interface.

This included the Structural Similarity Index Measurement (SSIM), Peak SNR (PSNR), Mean Absolute Error (MAE), and Normalized Mean Squared Error (NMSE). The evaluation for each metric was calculated using masks defined by the organizers to reduce potential bias from outside intensities. A final score was computed from the metrics:

$$FinalScore = \lambda_4(SSIM) + \lambda_5(\frac{PSNR}{32}) + \lambda_5(1 - MAE) + \lambda_5(1 - NMSE) \quad (3)$$

where $\lambda_4 = 0.7$ and $\lambda_5 = 0.1$.

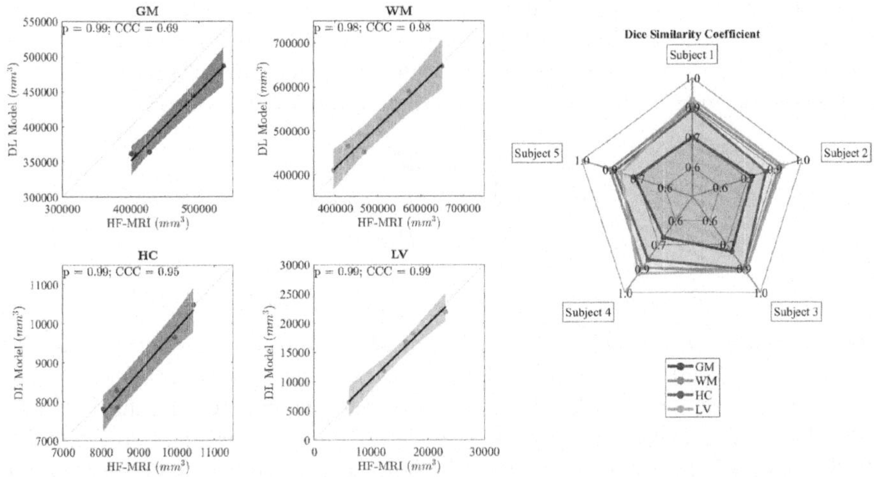

Fig. 2. Brain volume metrics including (left) scatterplots with Pearson correlation and Lin's CCC, and (right) a radar plot detailing the DSC for each of the evaluated subjects. Metrics were computed for the GM, WM, HC, and LV.

3 Results

Exploratory volumetric performance was assessed using the 5 validation cases from initial model fine-tuning (Fig. 2). We observed *very strong* correlations ($\rho >$ 0.8) for GM, WM, HC, and LV. Lin's CCC volumetric agreement was *excellent* (>0.9) for WM, HC, and LV, and *moderate* (0.5–0.75) for GM. The DSC scores were $0.74 \pm .03$ for GM, $0.88 \pm .01$ for WM, $0.84 \pm .02$ for HC, and $0.88 \pm .04$ for LV.

Performance on the ULF-EnC benchmark was evaluated on the hold out validation dataset of 10 subjects and testing dataset of 15 subjects (Table 2). Validation scoring returned 0.692 for SSIM, 28.641 for PSNR, 0.074 for MAE, and 0.068 for NMSE. The scoring on the testing dataset was overall lower than for validation with an SSIM of 0.685, PSNR of 28.463, MAE of 0.075, and 0.072 for NMSE.

Table 2. Model performance on the hold-out validation and testing datasets using metrics of the ULF-EnC. SSIM, PSNR, NMSE, MAE were automatically calculated using the Synapse Challenge interface.

	SSIM	PSNR	MAE	NMSE	Final Score
Validation	0.692	28.641	0.074	0.068	0.760
Testing	0.685	28.463	0.075	0.072	0.754

4 Discussion

DL-based enhancement methods have the potential to significantly improve brain volume analysis derived from ULF-MRIs, opening the door for routine examinations for brain health. Initiatives such as "brain charts" are already taking steps towards the creation of population-level brain volume growth curves across the lifespan, providing reference points for healthy aging or pathological onset [44]. ULF imaging combined with effective and accessible DL tools can help bridge the current gaps in representation for underserved populations. Here, the ULF-EnC marks an important step toward collaborative data sharing which can help further this goal. We aim to support such endeavors and aid in the discovery of novel strategies for improving ULF-MRI utility by proposing a relatively efficient 3D vision transformer model that utilizes both synthetic and real MRI training.

We designed our approach to optimize downstream brain volume analysis from enhanced ULF images, leveraging contrast information from axially-acquired FLAIR, T1w, and T2w ULF-MRIs. This utilized both mask-based and label-specific loss functions to constrain DL training within the brain and better preserve subcortical structures, respectively. FastSurfer was utilized for preliminary volumetric evaluation as a significantly faster alternative to FreeSurfer, the gold standard for T1w HF-MRI segmentation [39,45]. This produced *very strong* correlations for GM, WM, HC, and LV. However, Lin's CCC and DSC revealed lower agreement for GM volumes and segmentations. This may stem from the larger ULF voxel size, field-dependent contrast effects that influence the GM/WM boundary appearance, and spatial deformations related to the ULF permanent magnet design [20,46–49]. Existing studies have indicated that combining multi-directional ULF-MRIs may better capture anatomical information for volumetric analysis by leveraging the overlapping *k-space* regions [18–20]. Incorporating this information, such as an axial FLAIR, sagittal T1w, and coronal T2w, as input is likely to substantially boost volumetric performance, though a larger and dedicated testing dataset is needed for a robust evaluation.

Our model achieved high final scores of 0.760 and 0.754 on the validation and testing datasets, respectively, supporting the use of our hybrid training strategy that combined synthetic and real data. While the exclusive use of synthetic training data risks "domain shift" errors, this can provide a strong foundation for learning expected image intensities before exposure to real MRIs. However, our synthetic training did not account for regions outside of the brain, limiting its generalizability. Removing the mask-based loss functions from our fine-tuning step resulted in a substantially lower final score on the validation set of 0.735, underscoring its importance for model training. Future work may benefit from a comprehensive ablation study to compare model performance with and without synthetic pre-training and/or mask-based loss functions. The use of alternative segmentation tools, such as FreeSurfer, may improve structural delineation for more accurate model training [45]. Additionally, utilizing *real world* MRI data for "pre-training" may better capture true image intensities and biological variability than a synthetic approach.

Large-scale deployment of ULF-MRI could be greatly facilitated by the availability of lightweight DL enhancement models that are capable of operating in a variety of clinical or research settings, including those with limited resources. These practical considerations were a primary focus during model development, which aimed to optimize the efficiency of an established high-performance MRI synthesis technique [37]. The resulting 3D model utilized approximately 4.5 GB of GPU memory during evaluation and achieved a ~4.5x reduction in total parameters compared to the original 2D implementation. This significantly outperformed the standard 3D U-Net model in our testing, while only having ~1.8x more parameters. Despite high volumetric and image quality benchmark performance, model capacity was relatively small. Increasing the number of feature maps or adding larger multi-head attention would likely enhance structural representation and improve performance. Here, reduced precision computation (FP16 vs. FP32) or patch-based processing may help balance computational efficiency with model accuracy.

We have demonstrated a hybrid deep learning approach for enhancing ULF-MRI using limited paired training data, solely derived from the ULF-EnC dataset. While the current benchmark metrics primarily focused on image quality, our findings highlight the value of including MRI-derived biomarkers such as brain volumetric metrics to assess clinical utility. Accurately preserving small subcortical structures remains a key challenge, emphasizing the importance of expanding ULF datasets to include more paired cases and additional acquisition planes. Overall, our results demonstrate the promise of DL-enhanced ULF-MRI for improving brain volume analysis and suggest that continued data sharing and model development could help bring routine, accessible neuroimaging closer to clinical and research use.

Acknowledgments. This work has grant support through the NYU-KAIST Global Innovation and Research Institute and NIH P41 EB017183. This work was performed under the rubric of the Center for Advanced Imaging Innovation and Research (CAI2R, http://www.cai2r.net), an NIBIB National Center for Biomedical Imaging and Bioengineering.

Disclosure of Interests. The authors declare no conflicts of interests.

References

1. Brown, R.C., et al.: Neurodegenerative diseases: an overview of environmental risk factors. Environ. Health Perspect. **113**, 1250 (2005). https://doi.org/10.1289/ehp.7567
2. World population prospects 2022: summary of results. United Nations, New York (2022)
3. Fleming, K.A., et al.: The Lancet commission on diagnostics: transforming access to diagnostics. Lancet **398**, 1997–2050 (2021). https://doi.org/10.1016/S0140-6736(21)00673-5
4. Fox, N.C., Schott, J.M.: Imaging cerebral atrophy: normal ageing to Alzheimer's disease. Lancet **363**, 392–394 (2004). https://doi.org/10.1016/S0140-6736(04)15441-X
5. Frenzel, S., et al.: A biomarker for Alzheimer's disease based on patterns of regional brain atrophy. Front. Psychiatry. **10**, 953 (2020). https://doi.org/10.3389/fpsyt.2019.00953
6. Higgins-Chen, A.T., et al.: Aging biomarkers and the brain. Semin. Cell Dev. Biol. **116**, 180–193 (2021). https://doi.org/10.1016/j.semcdb.2021.01.003
7. Webb, A.G.: Magnetic resonance technology: hardware and system component design. Roy. Soc. Chem. (2016)
8. Geethanath, S., Vaughan, J.T., Jr.: Accessible magnetic resonance imaging: a review. J. Magn. Reson. Imaging **49**, e65–e77 (2019). https://doi.org/10.1002/jmri.26638
9. Jalloul, M., et al.: MRI scarcity in low- and middle-income countries. NMR Biomed. **36**, e5022 (2023). https://doi.org/10.1002/nbm.5022
10. van Dyck, C.H., et al.: Lecanemab in early Alzheimer's disease. N. Engl. J. Med. **388**, 9–21 (2023). https://doi.org/10.1056/NEJMoa2212948
11. de Vos, B., et al.: Design, characterisation and performance of an improved portable and sustainable low-field MRI system. Front. Phys. **9** (2021). https://doi.org/10.3389/fphy.2021.701157
12. Liu, Y., et al.: A low-cost and shielding-free ultra-low-field brain MRI scanner. Nat. Commun. **12**, 7238 (2021). https://doi.org/10.1038/s41467-021-27317-1
13. Prabhat, A.M., et al.: Methodology for low-field, portable magnetic resonance neuroimaging at the bedside. Front. Neurol. **12** (2021). https://doi.org/10.3389/fneur.2021.760321
14. Chetcuti, K., et al.: Implementation of a low-field portable MRI scanner in a resource-constrained environment: our experience in Malawi. AJNR Am. J. Neuroradiol. **43**, 670–674 (2022). https://doi.org/10.3174/ajnr.A7494
15. Parasuram, N.R., et al.: Future of neurology & technology: neuroimaging made accessible using low-field. Portable MRI Neurol. **100**, 1067–1071 (2023). https://doi.org/10.1212/WNL.0000000000207074
16. Abate, F., et al.: UNITY: a low-field magnetic resonance neuroimaging initiative to characterize neurodevelopment in low and middle-income settings. Dev. Cogn. Neurosci. **69**, 101397 (2024). https://doi.org/10.1016/j.dcn.2024.101397
17. Wu, E.X., Feng, X.: Reinventing magnetic resonance imaging for accessible healthcare: whole-body imaging at 0.05 Tesla. Clin. Transl. Med. **14**, e70071 (2024). https://doi.org/10.1002/ctm2.70071
18. Deoni, S.C.L., et al.: Simultaneous high-resolution T2-weighted imaging and quantitative T2 mapping at low magnetic field strengths using a multiple TE and multi-orientation acquisition approach. Magn. Reson. Med. **88**, 1273–1281 (2022). https://doi.org/10.1002/mrm.29273

19. Váša, F., et al.: Ultra-low-field brain MRI morphometry: test-retest reliability and correspondence to high-field MRI (2024). https://doi.org/10.1101/2024.08.14.607942
20. Hsu, P., et al.: Morphological brain analysis using ultra low-field MRI. Hum. Brain Mapp. **46**, e70232 (2025). https://doi.org/10.1002/hbm.70232
21. Pham, C.-H., et al.: Brain MRI super-resolution using deep 3D convolutional networks. In: 2017 IEEE 14th International Symposium on Biomedical Imaging (ISBI 2017), pp. 197–200 (2017). https://doi.org/10.1109/ISBI.2017.7950500
22. Preetha, C.J., et al.: Deep-learning-based synthesis of post-contrast T1-weighted MRI for tumour response assessment in neuro-oncology: a multicentre, retrospective cohort study. Lancet Digit. Health. **3**, e784–e794 (2021). https://doi.org/10.1016/S2589-7500(21)00205-3
23. Iglesias, J.E., et al.: SynthSR: a public AI tool to turn heterogeneous clinical brain scans into high-resolution T1-weighted images for 3D morphometry. Sci. Adv. **9**, eadd3607 (2023). https://doi.org/10.1126/sciadv.add3607
24. Billot, B., et al.: SynthSeg: segmentation of brain MRI scans of any contrast and resolution without retraining. Med. Image Anal. **86**, 102789 (2023). https://doi.org/10.1016/j.media.2023.102789
25. Lin, H., et al.: Low-field magnetic resonance image enhancement via stochastic image quality transfer. Med. Image Anal. **87**, 102807 (2023). https://doi.org/10.1016/j.media.2023.102807
26. Lau, V., et al.: Pushing the limits of low-cost ultralow-field MRI by dual-acquisition deep learning 3D superresolution. Magn. Reson. Med. (2023). https://doi.org/10.1002/mrm.29642
27. Chen, R.J., et al.: Synthetic data in machine learning for medicine and healthcare. Nat. Biomed. Eng. **5**, 493–497 (2021). https://doi.org/10.1038/s41551-021-00751-8
28. Kimberly, W.T., et al.: Brain imaging with portable low-field MRI. Nat. Rev. Bioeng. **1**, 617–630 (2023). https://doi.org/10.1038/s44222-023-00086-w
29. Lucas, A., et al.: Multi-contrast high-field quality image synthesis for portable low-field MRI using generative adversarial networks and paired data. medRxiv. 2023.12.28.23300409 (2023). https://doi.org/10.1101/2023.12.28.23300409
30. Lucas, A., et al.: Multisequence 3-T image synthesis from 64-mT low-field-strength MRI using generative adversarial networks in multiple sclerosis. Radiology **315**, e233529 (2025). https://doi.org/10.1148/radiol.233529
31. Islam, K.T., et al.: Improving portable low-field MRI image quality through image-to-image translation using paired low- and high-field images. Sci. Rep. **13**, 21183 (2023). https://doi.org/10.1038/s41598-023-48438-1
32. Dayarathna, S., et al.: Ultra low-field to high-field mri translation using adversarial diffusion. In: 2024 IEEE International Symposium on Biomedical Imaging (ISBI), pp. 1–4 (2024). https://doi.org/10.1038/s41598-023-48438-1
33. Islam, K.T., et al.: Ultra-low-field MRI image enhancement challenge (ULF-EnC) (2025). https://www.synapse.org/Synapse:syn65485242/wiki/631224
34. Dayarathna, S., et al.: Deep learning based synthesis of MRI, CT and PET: review and analysis. Med. Image Anal. **92**, 103046 (2024). https://doi.org/10.1016/j.media.2023.103046
35. Dayarathna, S., et al.: McCaD: multi-contrast MRI conditioned, adaptive adversarial diffusion model for high-fidelity MRI synthesis. In: 2025 IEEE/CVF Winter Conference on Applications of Computer Vision (WACV), pp. 670–679 (2025). https://doi.org/10.1109/WACV61041.2025.00075

36. Billot, B., et al.: Robust machine learning segmentation for large-scale analysis of heterogeneous clinical brain MRI datasets. Proc. Natl. Acad. Sci. **120**, e2216399120 (2023). https://doi.org/10.1073/pnas.2216399120
37. Dalmaz, O., et al.: ResViT: residual vision transformers for multi-modal medical image synthesis. IEEE Trans. Med. Imaging **41**, 2598–2614 (2022). https://doi.org/10.1109/TMI.2022.3167808
38. Pérez-García, F., et al.: TorchIO: a Python library for efficient loading, preprocessing, augmentation and patch-based sampling of medical images in deep learning. Comput. Methods Programs Biomed. **208**, 106236 (2021). https://doi.org/10.1016/j.cmpb.2021.106236
39. Henschel, L., Conjeti, S., Estrada, S., Diers, K., Fischl, B., Reuter, M.: FastSurfer - a fast and accurate deep learning based neuroimaging pipeline. Neuroimage **219**, 117012 (2020). https://doi.org/10.1016/j.neuroimage.2020.117012
40. McBride, G.: A proposal for strength-of-agreement criteria for Lin's concordance correlation coefficient. NIWA Client Rep. HAM2005-062, vol. 45, pp. 307–310 (2005)
41. Akoglu, H.: User's guide to correlation coefficients. Turk. J. Emerg. Med. **18**, 91–93 (2018). https://doi.org/10.1016/j.tjem.2018.08.001
42. Koo, T.K., Li, M.Y.: A guideline of selecting and reporting intraclass correlation coefficients for reliability research. J. Chiropr. Med. **15**, 155–163 (2016). https://doi.org/10.1016/j.jcm.2016.02.012
43. Zou, K.H., et al.: Statistical validation of image segmentation quality based on a spatial overlap index. Acad. Radiol. **11**, 178–189 (2004). https://doi.org/10.1016/S1076-6332(03)00671-8
44. Bethlehem, R.A.I., et al.: Brain charts for the human lifespan. Nature **604**, 525–533 (2022). https://doi.org/10.1038/s41586-022-04554-y
45. Fischl, B.: FreeSurfer. Neuroimage **62**, 774–781 (2012). https://doi.org/10.1016/j.neuroimage.2012.01.021
46. Natu, V.S., et al.: Apparent thinning of human visual cortex during childhood is associated with myelination. Proc. Natl. Acad. Sci. **116**, 20750–20759 (2019). https://doi.org/10.1073/pnas.1904931116
47. Mezer, A., et al.: Quantifying the local tissue volume and composition in individual brains with MRI. Nat. Med. **19**, 1667–1672 (2013). https://doi.org/10.1038/nm.3390
48. Ren, Z.H., et al.: Design and optimization of a ring-pair permanent magnet array for head imaging in a low-field portable MRI system. IEEE Trans. Magn. **55**, 1–8 (2019). https://doi.org/10.1109/TMAG.2018.2876679
49. Manso Jimeno, M., et al.: Superconducting magnet designs and MRI accessibility: a review. NMR Biomed. **36**, e4921 (2023). https://doi.org/10.1002/nbm.4921

Ultra-Low-Field Brain MRI Enhancement Using a Slice-Based Vision Transformer

Peter Hsu[1,2(✉)], Jeongsol Kim[3], Daniel Sodickson[1,2,4], Jong Chul Ye[3], Patricia Johnson[1,2,4], and Jelle Veraart[1,4]

[1] Bernard and Irene Schwartz Center for Biomedical Imaging, Department of Radiology, New York University Grossman School of Medicine, New York, NY, USA
peter.hsu@nyulangone.org

[2] Vilcek Institute of Graduate Biomedical Sciences, New York University Grossman School of Medicine, New York, NY, USA

[3] Kim Jaechul Graduate School of AI, Korea Advanced Institute of Science and Technology (KAIST), Seoul 06764, Republic of Korea

[4] Center for Advanced Imaging Innovation and Research (CAI2R), Department of Radiology, New York University Grossman School of Medicine, New York, NY, USA

Abstract. MRI is an indispensable technique for studying and monitoring brain health but remains inaccessible due to high costs, infrastructure requirements, and contraindications associated with conventional high-field (HF) systems. Ultra-low-field (ULF) MRI has emerged as a compelling cost-effective point-of-care neuroimaging solution with fewer contraindications, enabling deployment in diverse clinical and research settings. However, ULF-MRI is limited by inherently low signal-to-noise ratio (SNR), distinct field-dependent contrast properties, and relatively long scan times, impacting image quality and interpretability. Deep learning (DL) methods for MR image enhancement are promising solutions to address these challenges, but are currently constrained by the lack of publicly available ULF data. The ULF MRI Enhancement Challenge was established to address this gap by providing the first open dataset dedicated to benchmarking ULF image enhancement techniques. In this work, we present a 2D vision transformer for slice-based image translation from 64mT ULF-MRIs to 3T HF-MRIs across T1w, T2w, and FLAIR contrasts. Our method achieves high performance on image quality benchmarks, demonstrating the potential of this strategy for improving ULF-MRI utility in both research and clinical applications.

Keywords: Ultra-Low-Field MRI · Image Enhancement · Vision Transformer

1 Introduction

MRI is an essential tool for noninvasively visualizing the human brain. It provides superior image quality of soft tissue over other modalities, such as CT, with the

Z. Chen et al. (Eds.): ULF-EnC 2025, LNCS 16293, pp. 152–159, 2026.
https://doi.org/10.1007/978-3-032-23344-8_17

ability to capture multiple contrasts and high spatial resolution without ionizing radiation. This has led to its increased use for diagnosis and prognosis of neurological, psychiatric, and neurodegenerative diseases. However, most existing clinical MRI systems, here referred to as high-field (HF) MRI (1.5T or 3T), have limited accessibility due to their high cost, strict infrastructure requirements, and risk of contraindications [1,2]. This has created large healthcare disparities and has raised questions about the generalizability of existing neuroimaging studies and clinical trials to the wider population [3–5]. As a result, more accessible neuroimaging solutions are needed to bridge these gaps for equity in healthcare and research.

Ultra-low-field (ULF) MRI (<0.1T) is a promising development in this area, offering significant reductions in cost, flexible deployment solutions for diverse environments, and much fewer contraindications compared to HF systems [6–10]. This enables ULF-MRI to deliver, for example, point-of-care imaging in local or remote clinics and supports more inclusive research recruitment strategies [11,12]. However, the ultra-low field strength of these scanners leads to inherently low signal-to-noise ratio (SNR), often requiring lower spatial resolution and longer scan times to recover signal [13]. Indeed, common ULF acquisitions involve anisotropic voxel sizes and scan times of ~6 min per sequence [7,9]. While these imaging features can visualize large pathology, overall ULF image interpretability remains a challenge, in part because the field-dependency of MRI contrasts, ultimately limiting clinical and research utility.

Deep learning (DL) is well-positioned to overcome these limitations with established techniques for enhancing MR image quality through super-resolution and/or contrast transfer [14,15]. However, these strategies have yet to be fully developed for ULF-MRI, in part due to the scarcity of publicly available ULF datasets. Instead, existing state-of-the-art strategies tend to rely on synthetic training data which may have limited translatability to clinical applications with complex biology and pathology [16–19]. Ongoing efforts to collect ULF data are slow and the data is oftentimes not widely shared, hindering effective model training and comparable benchmarking across techniques. This has limited the translation of emerging strategies that have been leveraging real MRI data for training their algorithms [20–23]. The ULF-MRI Enhancement Challenge (ULF-EnC) aims to address these barriers by establishing the first open dataset designed for benchmarking ULF image enhancement methods [22–26].

Here, we propose a 2D vision transformer to perform slice-based image translation from axially-acquired 64mT ULF-MRIs to 3T HF-MRIs for T1w, T2w, and FLAIR imaging contrasts simultaneously. This approach is effective for ULF image enhancement, achieving high performance based on the image quality metrics of the challenge benchmark.

2 Methods

2.1 MRI Data

The ULF-EnC supplied a curated dataset for DL training and evaluation. This consisted of ULF- (64mT; Hyperfine Swoop) and HF-MRIs (3T; Siemens Bio-

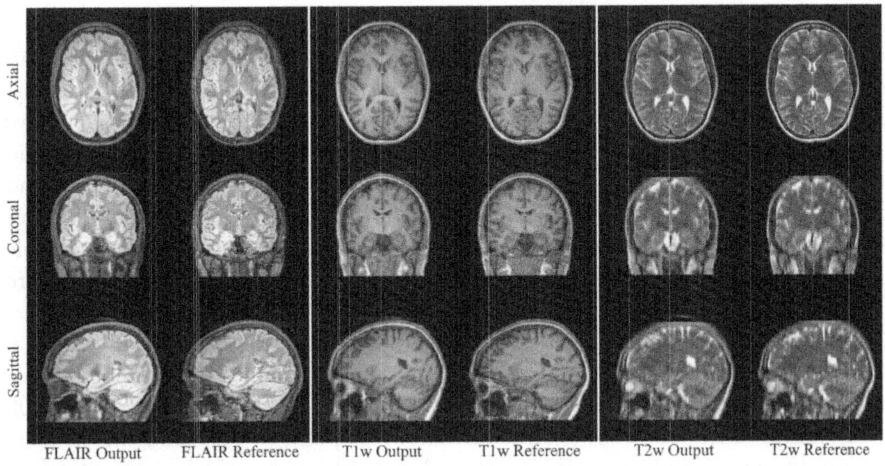

Fig. 1. Qualitative comparison between our proposed model output and the HF-MRI reference across FLAIR, T1w, and T2w contrasts.

graph mMR) from 50 training, 10 hold-out validation, and 15 testing subjects with FLAIR, T1w, and T2w contrasts. ULF images were acquired in the axial imaging direction with in-plane resolutions of $1.7 \times 1.7\,\text{mm}^2$ for FLAIR, $1.6 \times 1.6\,\text{mm}^2$ for T1w, and $1.5 \times 1.5\,\text{mm}^2$ for T2w, all with 5.0 mm slice thickness. FLAIR and T1w HF images were acquired with $1\,\text{mm}^3$ resolution, while T2w HF images were acquired in the axial imaging direction with $0.86 \times 0.86 \times 4.0\,\text{mm}^3$ resolution.

These images underwent the following preprocessing steps: 1mm^3 resolution resampling, bias field correction with FSL-FAST, and co-registration with FLIRT. Each image had dimensions of $224 \times 224 \times 160$, leading to 160 2D image slices per subject and 8000 total slices for DL model training. Each image slice was padded to 256×256 before being processed by our model and was cropped to 224×224 for evaluation.

2.2 Model Architecture and Training

A 2D vision transformer model architecture was adopted for ULF enhancement [27]. This enabled fuller utilization of the training dataset as each slice was processed as an individual image. Axial slices were selected for model processing as they had the highest resolution and contained the most informative structural information. Input and output consisted of 3-channels denoting unique imaging contrasts to leverage complementary representations of brain structure. All model code was developed using Pytorch.

The selected model architecture includes an encoder and decoder structure with feature maps of [64, 128, 256] and [256, 128, 64], respectively. This also utilizes transformer blocks that operate on 16×16 patches which are projected to a 768-dimensional embedding vector. These are then processed with multi-head self attention consisting of 12 attention heads with 64 dimensions per head.

The model was trained for 800 epochs using 45 ULF- and HF-MRIs from the training dataset. This utilized a combination of L1 and SSIM loss with 0.8 and 0.2 as their respective weightings. The AdamW optimizer (lr $= 1e^{-4}$) was applied with a batch size of 8 and a manual learning rate decay ($1e^{-8}$) was implemented for the final 400 epochs of training.

Initial model weight convergence was determined by 5 randomly selected "validation" subjects to avoid overfitting. To maximize full training dataset usage, the initial convergence weights were fine-tuned on the 5 "validation" cases for 10 epochs before evaluation on the hold-out validation and testing datasets.

2.3 Image Augmentation

Image augmentation was implemented using torchvision for model training which was applied on the original 224 × 224 image size. This included random rotation [−30, 30], random affine transformation (degrees = [−30, 30], translation = [0.1, 0.3], scale ranges = [0.8, 1.2], shears = [−10, 10]), Gaussian blurring, random image resizing and cropping, random Gaussian noise, random flipping, random sharpness, random gamma, and random elastic deformation.

2.4 Evaluation Metrics

Model performance was determined using the Structural Similarity Index Measurement (SSIM), Peak SNR (PSNR), Mean Absolute Error (MAE), Normalized Mean Squared Error (NMSE), and a combined weighted final score (0.7, 0.1, 0.1, and 0.1, respective weighting). These were calculated within the brain

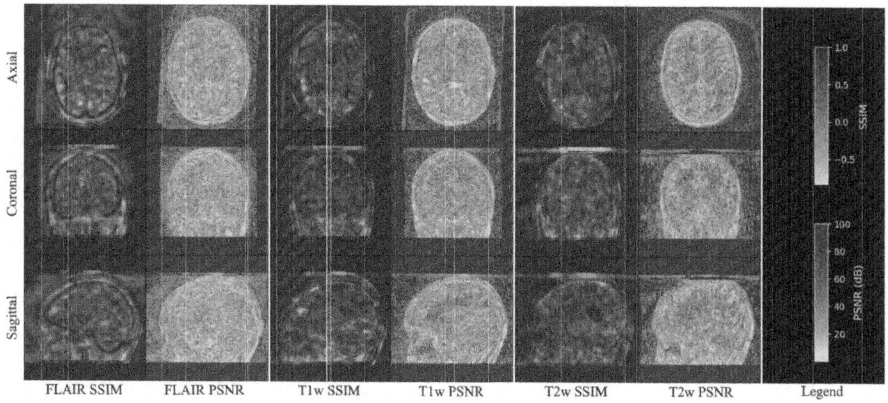

Fig. 2. Quantitative SSIM and PSNR maps comparing model output to the HF reference images across FLAIR, T1w, and T2w contrasts. Images were zero padded prior to this evaluation, leading to "null regions" with high similarity.

and extracerebral tissues through challenge-defined image masking and were computed automatically via the Synapse web interface where the challenge was hosted.

3 Results

Model output image quality, prior to fine-tuning, compared to the target HF images is presented in Fig. 1. We further created quantitative maps to identify local performance based on the SSIM and PSNR metrics (Fig. 2).

Performance on the challenge benchmark is showcased in Table 1. Our approach led to strong performance on the hold out validation set with a SSIM of 0.668, PSNR of 29.578, MAE of 0.76, and NMSE of 0.078. Performance on the testing set was similar, with a SSIM of 0.669, PSNR of 29.457, MAE of 0.076 and NMSE or 0.078.

Table 1. Performance metrics for our model based on the ULF EnC Challenge hold-out validation dataset. This includes the SSIM, PSNR, MAE, and NMSE comparing whole-image output and masked image output

	SSIM	PSNR	MAE	NMSE	Final Score
Validation	0.668	29.578	0.076	0.078	0.745
Testing	0.669	29.457	0.076	0.078	0.745

4 Discussion

The availability of paired ULF- and HF-MRIs substantially benefits the development of DL models for image enhancement. While previous approaches had to approximate ULF imaging features with synthetic training data, *real world* images inherently encode true biological and scanner variability, leading to more robust enhancement properties [17–19]. Our proposed model framework simultaneously improves the image quality of T1w, T2w, and FLAIR ULF-MRIs to leverage tissue representation across multiple contrasts. This is in contrast to previous methods trained on real data which developed separate models to enhance each individual contrast [22,23].

Our results show strong qualitative correspondence between model output and target HF data. This is supported quantitatively by high SSIM (0.669) and PSNR (29.457) performance on the testing dataset. Interestingly, our model achieved the same final score across both validation and testing datasets, indicating strong consistency in image quality enhancement performance. Our exploratory quantitative maps indicate that the largest SSIM and PSNR discrepancies occur in extracerebral tissues, such as the skull, which are typically outside the primary focus of neuroimaging studies. This could be addressed by focusing model training or evaluation to skull-stripped images, which may be more indicative of potential research utility. We also observed some small discrepancies for regions of lower GM SSIM which may be attributed to the large field strength difference between input and target images influencing the GM/WM boundary appearance and challenging accurate enhancement [28,29]. This could be further assessed through cortical segmentation or thickness evaluations between model output and HF reference data.

A 2D model architecture was selected to maximize training examples as image slices from a relatively small dataset of 50 subjects. A transformer was selected over other model configurations for their inherently improved spatial information encoding which is beneficial for image translation tasks. However, this led to a limited ability to account for inter-slice information. As a result, imaging artifacts were observed near the top of the head and skull, where discontinuities between slices across contrasts were more apparent. These introduced noticeable image discrepancies that negatively impacted model performance. Examples are visible in the coronal and sagittal planes of our quantitative maps (Fig. 2). While image thresholding was applied in an attempt to reduce these artifacts, it was ultimately too inconsistent across our testing. Future work may benefit from a series of single-channel models that each process a unique imaging contrast or transitioning to a 3D model architecture to reduce such artifacts. Alternative model architectures such as GANs or diffusion models may also further improve performance.

Future iterations of this challenge may benefit from reporting confidence intervals for all metrics to better assess statistical significance between models. Including segmentation-derived measures, such as dice similarity coefficient, would also provide valuable insights into brain structure correspondence between model output and target HF-MRI. Including ULF cases with pathology may offer unique out-of-distribution testing, enabling a robust evaluation of model generalizability and consistency between inputs and outputs. Expanding acquisitions beyond axial scans to include coronal and sagittal views could further enhance tissue characterization by providing complementary anatomical information. Such additions may lead to novel model architectures designed to integrate multiple FoVs for improved enhancement.

Altogether, our proposed DL method showcases effective ULF image enhancement through high structural similarity with HF-MRI. This is promising for improving the clinical and research utility of ULF-MRI, potentially improving clinical readability and enabling research compatibility with established workflows.

Acknowledgments. This work has grant support through the NYU-KAIST Global Innovation and Research Institute and NIH P41 EB017183. This work was performed under the rubric of the Center for Advanced Imaging Innovation and Research (CAI2R, www.cai2r.net), an NIBIB National Center for Biomedical Imaging and Bioengineering.

Disclosure of Interests. The authors declare no conflicts of interests.

References

1. Sarracanie, M., LaPierre, C.D., Salameh, N., Waddington, D.E.J., Witzel, T., Rosen, M.S.: Low-cost high-performance MRI. Sci. Rep. **5**, 15177 (2015). https://doi.org/10.1038/srep15177
2. Ghadimi, M., Sapra, A.: Magnetic resonance imaging contraindications. In: StatPearls. StatPearls Publishing, Treasure Island (FL) (2025)
3. Hilabi, B.S., Alghamdi, S.A., Almanaa, M.: Impact of magnetic resonance imaging on healthcare in low- and middle-income countries. Cureus **15**, e37698 (2023). https://doi.org/10.7759/cureus.37698
4. Jalloul, M., et al.: MRI scarcity in low- and middle-income countries. NMR Biomed. **36**, e5022 (2023). https://doi.org/10.1002/nbm.5022
5. van Dyck, C.H., et al.: Lecanemab in early alzheimer's disease. N. Engl. J. Med. **388**, 9–21 (2023). https://doi.org/10.1056/NEJMoa2212948
6. Liu, Y., et al.: A low-cost and shielding-free ultra-low-field brain MRI scanner. Nat. Commun. **12**, 7238 (2021). https://doi.org/10.1038/s41467-021-27317-1
7. Chetcuti, K., et al.: Implementation of a low-field portable MRI scanner in a resource-constrained environment: our experience in Malawi. AJNR Am. J. Neuroradiol. **43**, 670–674 (2022). https://doi.org/10.3174/ajnr.A7494
8. Guallart-Naval, T., et al.: Portable magnetic resonance imaging of patients indoors, outdoors and at home. Sci. Rep. **12**, 13147 (2022). https://doi.org/10.1038/s41598-022-17472-w
9. Parasuram, N.R., et al.: Future of neurology amp; technology: neuroimaging made accessible using low-field. Portable MRI. Neurology. **100**, 1067–1071 (2023). https://doi.org/10.1212/WNL.0000000000207074
10. de Vos, B., Remis, R.F., Webb, A.G.: An integrated target field framework for point-of-care halbach array low-field MRI system design. Magn. Reson. Mater. Phys., Biol. Med. **36**, 395–408 (2023). https://doi.org/10.1007/s10334-023-01093-z
11. Mazurek, M.H., et al.: Portable, bedside, low-field magnetic resonance imaging for evaluation of intracerebral hemorrhage. Nat. Commun. **12**, 5119 (2021). https://doi.org/10.1038/s41467-021-25441-6
12. Deoni, S.C.L., et al.: Development of a mobile low-field MRI scanner. Sci. Rep. **12**, 5690 (2022). https://doi.org/10.1038/s41598-022-09760-2
13. Kimberly, W.T., et al.: Brain imaging with portable low-field MRI. Nat. Rev. Bioeng. **1**, 617–630 (2023). https://doi.org/10.1038/s44222-023-00086-w
14. Pham, C.-H., Ducournau, A., Fablet, R., Rousseau, F.: Brain MRI super-resolution using deep 3D convolutional networks. In: 2017 IEEE 14th International Symposium on Biomedical Imaging (ISBI 2017), pp. 197–200 (2017). https://doi.org/10.1109/ISBI.2017.7950500
15. Preetha, C.J., et al.: Deep-learning-based synthesis of post-contrast T1-weighted MRI for tumour response assessment in neuro-oncology: a multicentre, retrospective cohort study. Lancet Digit. Health. **3**, e784–e794 (2021). https://doi.org/10.1016/S2589-7500(21)00205-3
16. Chen, R.J., Lu, M.Y., Chen, T.Y., Williamson, D.F.K., Mahmood, F.: Synthetic data in machine learning for medicine and healthcare. Nat. Biomed. Eng. **5**, 493–497 (2021). https://doi.org/10.1038/s41551-021-00751-8
17. Iglesias, J.E., et al.: SynthSR: a public AI tool to turn heterogeneous clinical brain scans into high-resolution T1-weighted images for 3D morphometry. Sci. Adv. **9**, eadd3607 (2023). https://doi.org/10.1126/sciadv.add3607

18. Lin, H., et al.: Low-field magnetic resonance image enhancement via stochastic image quality transfer. Med. Image Anal. **87**, 102807 (2023). https://doi.org/10.1016/j.media.2023.102807
19. Lau, V., et al.: Pushing the limits of low-cost ultralow-field MRI by dual-acquisition deep learning 3D superresolution. Magn. Reson. Med. (2023). https://doi.org/10.1002/mrm.29642
20. Lucas, A., et al.: Multi-contrast high-field quality image synthesis for portable low-field MRI using generative adversarial networks and paired data. medRxiv. 2023.12.28.23300409 (2023). https://doi.org/10.1101/2023.12.28.23300409
21. Lucas, A., et al.: Multisequence 3-T image synthesis from 64-mT low-field-strength MRI using generative adversarial networks in multiple sclerosis. Radiology **315**, e233529 (2025). https://doi.org/10.1148/radiol.233529
22. Islam, K.T., et al.: Improving portable low-field MRI image quality through image-to-image translation using paired low- and high-field images. Sci. Rep. **13**, 21183 (2023). https://doi.org/10.1038/s41598-023-48438-1
23. Dayarathna, S., Islam, K.T., Chen, Z.: Ultra low-field to high-field MRI translation using adversarial diffusion. In: 2024 IEEE International Symposium on Biomedical Imaging (ISBI), pp. 1–4 (2024). https://doi.org/10.1109/ISBI56570.2024.10635808
24. Islam, K.T., Peiris, H., Zhonz, S., Dayarathna, S., Zakavi, P., Chen, Z.: Ultra-low-field MRI image enhancement challenge (ULF-EnC) (2025). https://www.synapse.org/Synapse:syn65485242/wiki/631224
25. Dayarathna, S., Islam, K.T., Uribe, S., Yang, G., Hayat, M., Chen, Z.: Deep learning based synthesis of MRI, CT and PET: review and analysis. Med. Image Anal. **92**, 103046 (2024). https://doi.org/10.1016/j.media.2023.103046
26. Dayarathna, S., et al.: McCaD: multi-contrast MRI conditioned, adaptive adversarial diffusion model for high-fidelity MRI synthesis. In: 2025 IEEE/CVF Winter Conference on Applications of Computer Vision (WACV), pp. 670–679 (2025). https://doi.org/10.1109/WACV61041.2025.00075
27. Dalmaz, O., Yurt, M., Çukur, T.: ResViT: residual vision transformers for multimodal medical image synthesis. IEEE Trans. Med. Imaging **41**, 2598–2614 (2022). https://doi.org/10.1109/TMI.2022.3167808
28. Mezer, A., et al.: Quantifying the local tissue volume and composition in individual brains with MRI. Nat. Med. **19**, 1667–1672 (2013). https://doi.org/10.1038/nm.3390
29. Natu, V.S., et al.: Apparent thinning of human visual cortex during childhood is associated with myelination. Proc. Natl. Acad. Sci. **116**, 20750–20759 (2019). https://doi.org/10.1073/pnas.1904931116

Multi-input Generalised-Hilbert Mamba for Super-Resolution of Ultra-Low-Field MRI

Levente Baljer[1,2(✉)], Niall Bourke[1], Zhenshan Xie[2], Emma Robinson[2], and František Váša[1]

[1] Department of Neuroimaging, King's College London, London, UK
levente.1.baljer@kcl.ac.uk
[2] Department of Biomedical Computing, King's College London, London, UK

Abstract. Brain magnetic resonance imaging (MRI) is essential for both clinical decision-making and research, yet access to high-field (HF) systems is hindered by high costs, infrastructure requirements and long waiting lists. Ultra-low-field (ULF) systems combined with deep-learning super-resolution (SR) methods offer a solution by enabling reconstruction of higher resolution images from low-resolution acquisitions. In this work, we build on a previous hybrid CNN- and state-space-model (SSM)-based architecture, GAMBAS, extending it to integrate multi-contrast input data. Our proposed model incorporates three separate encoders for T_1-weighted, T_2-weighted, and FLAIR images, fusing their latent representations into a shared bottleneck. This architecture enables the model to leverage complementary information across modalities to produce high-quality, high-resolution outputs of all three contrasts. We demonstrate the method using an adult paired ULF and HF neuroimaging dataset, highlighting its potential to improve imaging workflows in diverse global healthcare and research settings.

Keywords: Super-resolution · Mamba · Multimodal imaging

1 Introduction

1.1 Ultra-Low-Field Imaging

High-field (HF) MRI (>1 T) is essential for non-invasive research and clinical assessment. However, super-conducting HF scanners are expensive (approximately $1,000,000 per Tesla [1]) and have extensive infrastructural requirements, resulting in pronounced disparities in global access. This, in turn, causes participants from low- and middle-income countries (LMICs) to be underrepresented in neuroimaging studies, and patients in these settings to experience significantly prolonged waiting times for clinical evaluation.

Ultra-low-field (ULF) MRI systems (<0.1 T) leveraging permanent magnets have the potential to reduce this access gap. These systems are substantially

Z. Chen et al. (Eds.): ULF-EnC 2025, LNCS 16293, pp. 160–172, 2026.
https://doi.org/10.1007/978-3-032-23344-8_18

cheaper to purchase and run [2], and their open design and quiet operation improve patient tolerance. The trade-off, however, is a marked reduction in resolution and signal-to-noise ratio, producing images that may be insufficient for confident clinical interpretation or for pre-processing by existing analysis pipelines optimised for HF scans.

1.2 Super-Resolution

Deep-learning-based super-resolution (SR) offers a potential remedy by reconstructing high-quality images from ULF acquisitions at minimal additional cost. The field of medical image translation and synthesis has historically relied on convolutional neural networks (CNNs) [3–5], with researchers often incorporating adversarial training to boost perceptual realism [6–8]. Although efficient in processing high-dimensional medical imaging data, CNNs lack the ability to encapsulate long-range spatial dependencies [9], essential for learning accurate feature representations of brain structures dispersed across both neighbouring and distant voxels (e.g. cortical grey matter or cerebrospinal fluid). As such, more recent work leverages architectures capable of modelling long-range context by supplementing CNNs with Vision Transformers (ViTs) [10,11] or State-Space-Models (SSMs) such as Mamba [12,13], with the latter of these two approaches demonstrating increased computational efficiency [14].

Many prior models have exhibited impressive SR of brain MRI within contrasts, e.g. by super-resolving an ULF T_2-weighted scan to approximate an HF T_2-weighted target. This includes our earlier work, GAMBAS (**G**eneralised-Hilbert **Mamba** **S**uper-resolution) [15], a hybrid CNN-Mamba architecture designed for SR of T_2-weighted paediatric ULF scans. However, several other groups have demonstrated that it is possible to leverage multi-contrast acquisition protocols, using T_1-weighted, T_2-weighted and/or FLAIR acquisitions [16,17] to guide the generation process. Here we introduce GAMBAS+ as an extension of GAMBAS, building upon prior work by using three contrast-specific encoder heads and a gating network to form a multi-contrast fused latent embedding. We train our model on a dataset of 50 adult participants, demonstrating significant improvements compared to the previous single-contrast model across diverse performance metrics.

2 Methods

2.1 Preliminaries

Mamba and SSMs: State space models (SSMs) are continuous-time latent state models, which map a 1D input $x(t) \in \mathbb{R}^L$ to an output $y(t) \in \mathbb{R}^L$ through hidden state $h(t) \in \mathbb{R}^N$ via the following linear ODE:

$$\begin{aligned} h'(t) &= \mathbf{A}h(t) + \mathbf{B}x(t), \\ y(t) &= \mathbf{C}h(t) + \mathbf{D}x(t) \end{aligned} \tag{1}$$

Here, $\mathbf{A} \in \mathbb{R}^{N \times N}$, $\mathbf{B} \in \mathbb{R}^{N \times 1}$, and $\mathbf{C} \in \mathbb{R}^{1 \times N}$ are learnable parameters and $\mathbf{D} \in \mathbb{R}^1$ represents a residual connection. The continuous-time SSM has prohibitive computation and memory requirements for deep learning applications, however, it can be discretized using a timescale parameter $\boldsymbol{\Delta}$. For this, the zero-order hold (ZOH) technique is employed, whereby continuous parameters $\mathbf{A}$, $\mathbf{B}$ are discretized as $\overline{\mathbf{A}} = \exp(\boldsymbol{\Delta}\mathbf{A})$, $\overline{\mathbf{B}} = (\boldsymbol{\Delta}\mathbf{A})^{-1}(\exp(\boldsymbol{\Delta}\mathbf{A}) - \mathbf{I}) \cdot \boldsymbol{\Delta}\mathbf{B}$. Substituting these variables into Eq. (1) yields the structured state-space sequence model, or S4 [18], represented in the following recurrent form:

$$\begin{aligned} h_t &= \overline{\mathbf{A}} h_{t-1} + \overline{\mathbf{B}} x_t, \\ y_t &= \mathbf{C} h_t + \mathbf{D} x_t \end{aligned} \tag{2}$$

Although efficient for training, S4 is unable to select data in an input-dependent manner. Mamba overcomes this hurdle by setting the model parameters to be input-dependent, allowing for a selection mechanism that can filter out irrelevant information and remember relevant information indefinitely. Crucially for medical imaging tasks with high-dimensional inputs, Mamba demonstrates an ability to model long-range dependencies on par with that of Transformers, while significantly reducing computational complexity [12].

Space-Filling Curves: As Mamba is designed to process 1D data, its application to vision tasks requires image features of shape (B, C, H, W, D) to be flattened and transposed to (B, L, C), where $L = H \times W \times D$. Unlike input sequences in natural language processing tasks, images have no inherent directionality. As such, the approach used to arrange voxels into a 1D sequence determines the way Mamba processes the image and, in turn, affects overall model performance [19]. Space-filling curves have gained increasing use in this domain, owing to their ability to provide a linear mapping between higher and lower dimensional spaces (e.g. 3D and 1D) in a manner that preserves spatial locality. Although several variations exist, the Hilbert curve demonstrates superior order-preserving behaviour compared to alternatives [20]. Consequently, we employ a variant of the Hilbert curve tailored to handle 3D inputs [21] to serialise structural MRI input into a 1D vector.

2.2 Architecture

Our architecture is based on GAMBAS [15]. It uses a 3D counterpart of the 9-block ResNet architecture [22], an encoder-decoder network with an extended central bottleneck (consisting of 9 layers). In the bottleneck, the first, middle, and end layers are supplemented with Mamba blocks, wherein 3D encoded features are serialised via a Generalised-Hilbert scan and fed into bidirectional Mamba modules. Inspired by earlier work, the outputs from Mamba are concatenated with CNN outputs and fed into channel-mixing blocks [23] to mix contextual and structural features from SSMs and CNNs, respectively. These blocks consist of two parallel convolutional branches of differing kernel sizes, mixing features at varying spatial locations while maintaining the dimensionality of the latent embedding.

To allow our model to handle multi-contrast input, we modified the architecture to have three separate encoder heads for each of the three MRI contrasts (T_1w, T_2w, and FLAIR images). Once each image is encoded, the three contrasts are stacked in the channel dimension, and fed into a gating network. In similar fashion to [24], we use the gating network to calculate weights for each of the three contrast embeddings from a single subject, allowing our model to form a fused latent embedding in an input-dependent manner. Importantly, the gating network is also formed of a combination of CNN and Mamba layers to allow for both local and global context to be encapsulated (Fig. 1). To view the relative contribution of each element of our model, see Appendix A. We maintain a single decoder for our model, such that three input contrasts are used to generate one of either T_1w, T_2w, or FLAIR outputs. As such, separate networks are trained for each output modality.

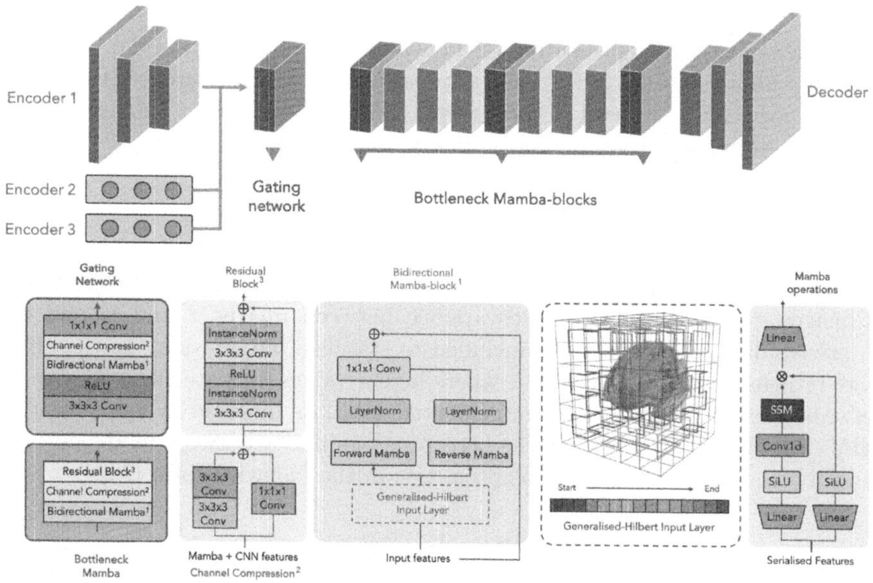

Fig. 1. GAMBAS+ Generator Architecture: Encoders 1, 2, and 3 process T_1w, T_2w and FLAIR, respectively, with the gating network (red) learning fusion weights using a bidirectional Mamba-block and channel compression module, inserted between two layers of CNNs. Bottleneck layers 1, 5 and 9 (blue) include forward and reverse Mamba blocks, channel compression and residual blocks, with remaining layers (green) comprising of residual blocks only.

2.3 Training Strategy

We use an adversarial training scheme to train each of our networks, implementing PatchGAN as our discriminator [25]. The discriminator has two input channels, taking in both an ULF scan and either a real or synthetic HF scan at each training step to determine whether the input pair is real. Although the generator receives three ULF input scans, the discriminator only receives the relevant ULF contrast (i.e. T_1w ULF for the T_1w model). As such, our model follows the conditional GAN (cGAN) training objective, summarised by:

$$\mathcal{L}_{cGAN}(G, D) = \mathbb{E}_{x,y}[\log D(x, y)] + \mathbb{E}_x[\log 1 - D(x, (G(x))] \tag{3}$$

where generator G and discriminator D aim to a minimise and maximise expectation $\mathbb{E}$, respectively. To ensure subject anatomy is maintained on conditional image generation, G is tasked with an additional loss of minimising the L1 distance between model outputs and reference HF images:

$$\mathcal{L}_{L1}(G) = \mathbb{E}_{x,y}[\|y - G(x)\|_1] \tag{4}$$

To enhance the quality of our model outputs, we additionally add a structural similarity index measure (SSIM) loss between synthetic and reference HF images. SSIM is normally composed of three terms (luminance, contrast, and structure), however a simplified, combined formulation is given below:

$$\mathcal{L}_{SSIM}(G) = 1 - \frac{(2\mu_x\mu_y + C_1)(2\sigma_{xy} + C_2)}{(\mu_x^2 + \mu_y^2 + C_1)(\sigma_x^2 + \sigma_y^2 + C_2)} \tag{5}$$

Here, μ_x, μ_y and σ_x^2, σ_y^2 are mean intensities and variances, respectively, of images x and y, σ_{xy} is the covariance between images x and y, and C_1, C_2 are small positive constants included to stabilise the division. As the SSIM score returns a value between 0–1, where higher values indicate closer alignment between images, this metric is implemented as a loss function by calculating 1-SSIM. Finally, to further boost the perceptual similarity between model outputs and target images, we add a perceptual loss in the following form [26]:

$$\mathcal{L}_{\text{Perc}}(G) = \mathbb{E}_{x,y}\left[\sum_{l \in \mathcal{L}} w_l \frac{1}{N_l} \parallel \phi_l(y) - \phi_l(G(x))\|_2^2\right] \tag{6}$$

Here, $\phi_l(G(x))$ and $\phi_l(y)$ denote activations of a fixed, pretrained CNN (in this case, a ResNet trained on the RadImageNet dataset [27]) at layer l, given synthetic HF scan x and real HF scan y as inputs. After computing an elementwise ℓ_2 norm between activations, the difference is spatially normalised via $N_l = C_l H_l W_l$ and summed across a set of chosen layers $\mathcal{L}$, with each individual layer weighted by pre-specified hyperparameter w_l. Both the SSIM and Perceptual loss are implemented using the MONAI PyTorch framework [28]. Combining Eq. (3)–Eq. (6) yields the following overall objective for our model:

$$\begin{aligned}\mathcal{L}_{\text{Total}} = & \arg\min_G \max_D \lambda_1 \mathcal{L}_{cGAN}(G, D) + \lambda_2 \mathcal{L}_{L1}(G) \\ & + \lambda_3 \mathcal{L}_{SSIM}(G) + \lambda_4 \mathcal{L}_{Perc}(G)\end{aligned} \tag{7}$$

The weighting terms for each loss are set at $\lambda_1 = 1$, $\lambda_2 = 300$, $\lambda_3 = 100$ and $\lambda_4 = 0.1$. These values were determined through hyperparameter tuning experiments outlined in Appendix B.

2.4 Model Prediction

To further enhance the accuracy and robustness of our outputs, we employ an ensemble model strategy. Ensemble models have primarily seen use in tasks that involve discrete outputs, e.g. object detection, semantic segmentation, and edge detection, where multiple model predictions are fused via techniques such as majority voting or probability averaging [29]. Application in image generation is often less straightforward, as fusion of outputs can lead to the loss of fine-grained structure. This has led researchers to seek alternative ensembling techniques, such as fusion of multiple generative models in the latent space [30]. Nevertheless, in Appendix C we demonstrate how, for our given image-translation task, fusion in image-space allows for areas of high uncertainty in ultra-low-field inputs to be corrected, and we further provide experiments on the optimal number of individual models. Our final approach involves training four identical models under differing random initializations for each image contrast, followed by voxel-wise averaging of model predictions.

2.5 Dataset

All MRI data used in this study were acquired at the Monash Biomedical Imaging facility in Monash University, Australia, and were provided as part of the Ultra-Low-Field MRI Image Enhancement Challenge (ULF-EnC) [31]. Brain scans from 75 adults were obtained using both a Siemens Biograph mMR 3T scanner (Erlangen, DE) and a Hyperfine Swoop 64mT system (Guildford, CT). On each scanner, three separate contrasts were acquired, yielding a total of 6 scans per subject. Scans on the 3T system included an isotropic T_1w sequence (1.0 × 1.0 × 1.0 mm), a non-isotropic T_2w sequence (0.86 × 0.86 × 4.0 mm), and an isotropic FLAIR sequence (1.0 × 1.0 × 1.0 mm). All scans on the 64mT system were acquired at non-isotropic resolutions, including a T_1w sequence (1.6 × 1.6 × 5.0 mm), a T_2w sequence (1.5 × 1.5 × 5.0 mm), and a FLAIR sequence (1.7 × 1.7 × 5.0 mm). Consequently, preprocessing involved resampling of all scans to a standardized isotropic resolution of $1\,\text{mm}^3$, resizing to a uniform matrix size of 224 × 224 × 160, and bias-field correction using FSL-FAST to mitigate intensity inhomogeneities. To enable paired training of super-resolution models, all subject-specific 64mT and 3T scans were co-registered using FSL FLIRT [32]. Furthermore, participant privacy was ensured via the application of an identical face removal mask to all scans.

2.6 Experiment Details

The data were split into training/validation/test sets of 50/10/15 scans (i.e. ~67/13/20%). We trained our models for 1000 epochs using the Adam optimizer with $\beta_1 = 0.5$, $\beta_2 = 0.999$ and a learning rate of 0.0002 that decayed linearly to 0 over the last 500 epochs. Patch-wise training was conducted, extracting random $128 \times 128 \times 128$ patches from input and target images at each training step, and batch size was set to 1. During training, image augmentation was carried out on the fly, including random 3D rotation, BSpline deformation, random flip, and contrast and brightness adjustment. All model training was carried out using NVIDIA A30 Tensor Cores (40GB) using King's College London's Computational Research, Engineering and Technology Environment (CREATE) [33].

2.7 Evaluation Metrics

As part of the ULF-EnC challenge, HF targets for the 10 validation and 15 test subjects were not available to us; instead, model outputs for these groups were uploaded onto the challenge portal for evaluation by the organisers. This evaluation involved the calculation of SSIM, peak signal-to-noise ratio (PSNR), mean absolute error (MAE) and normalised root mean squared error (NRMSE) between synthetic outputs and target images. To ensure that only voxels including brain tissue and skull contributed to the metrics, a mask was applied to all scans to eliminate background voxels. Each of the four metrics were averaged across contrasts, and a final score was additionally computed based on the following weightings: 0.7 × SSIM + 0.1 × (PSNR/32) + 0.1 × (1-MAE) + 0.1 × (1-NRSME). To compare approaches, these values were computed for raw ULF scans, outputs from GAMBAS, and outputs from GAMBAS+. Importantly, only the final best model was evaluated on the test set, whereas other models were evaluated on the validation set.

3 Results

From our comparative analyses, we demonstrate that GAMBAS+ significantly enhances the visual quality of MRI scans compared to input ULF images (Fig. 2). We highlight this with improved performance in all voxelwise metrics (Table 1),

Table 1. Performance of GAMBAS+ averaged across contrasts, compared to GAMBAS and raw ULF scans.

Model	SSIM (↑)	PSNR (↑)	MAE (↓)	NMSE (↓)	Score (↑)
None (ULF)	0.427	21.376	0.151	0.541	0.497
GAMBAS	0.665	29.119	0.074	0.179	0.731
GAMBAS+	0.729	30.070	0.066	0.061	0.792

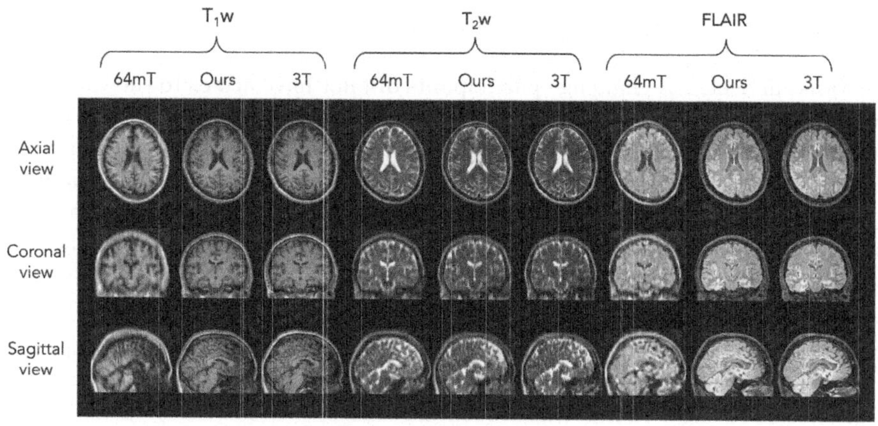

Fig. 2. MRI slices from a single test subject. Left to right: raw ULF scan, GAMBAS+ output and reference HF scan, sorted according to contrast (T_1w, T_2w and FLAIR).

with GAMBAS+ improving ULF scores on average by 64.1%. We further demonstrate improved performance relative to the original, single-contrast GAMBAS model, with voxelwise metrics rising by an average of 22.4% (Table 1).

4 Discussion

Our network, GAMBAS+, presents an effective and reliable method for translation of multi-contrast ULF to single-contrast HF MRI, highlighting the potential for deep learning tools in widening accessibility to MRI. By endowing our model with a gating network, which itself contains a Generalised-Hilbert Mamba block, we allow the model to learn dynamic weighting matrices for the inputs, in a manner that takes into account long-range spatial dependencies among voxels. Using this method, we demonstrate improved performance compared to our earlier single-contrast model, underscoring the benefits conferred by the formation of a fused latent representation of three contrasts to guide image synthesis.

One notable limitation of our method is that we do not generate multiple output contrasts with a single model. With the method we outline, three separate models must be trained to be able to produce T_1w, T_2w, and FLAIR outputs, incurring significantly greater computational costs. We opted for this approach to maximise accuracy of outputs by simplifying the generation task for each of our models. Nevertheless, future work will explore alternative, more efficient synthesis strategies.

Acknowledgments. We would like to acknowledge contributions from Prof. Rosalyn Moran, Prof. Steven Williams, and Prof. Sean Deoni, whose combined support allowed for the construction of our original GAMBAS model.

Disclosure of Interests. The authors have no competing interests to declare that are relevant to the content of this article.

A Model Architecture Ablations

As stated in Sect. 2.7, challenge participants did not have access to the validation or test sets. As such, all experiments listed here were carried out by splitting our training set of 50 subjects into training/validation sets of 40/10 scans (i.e. 80/20%), and calculating image-quality metrics using the 10 internal validation subjects. For brevity, only T_1w models were trained and no masking was applied (i.e. full-volume T_1w outputs were compared to full-volume T_1w HF targets) (Table 2).

Table 2. Performance of a baseline CNN (9-block ResNet) compared to variants with the modular addition of generalised-Hilbert Mamba blocks (Mamba), Channel-mixing (CM) blocks, Gating network with only convolutional operaotors (GN^{CNN}), and Gating network with convolutional and SSM operators (GN^{Mamba}).

Model	SSIM (↑)	PSNR (↑)	MAE (↓)	NMSE (↓)	Score (↑)
CNN [22]	0.824	22.295	0.0337	0.0938	0.834
+ Mamba	0.792	21.787	0.0367	0.1066	0.808
+ CM block	0.827	22.285	0.0335	0.0939	0.836
+ GN^{CNN}	0.830	22.608	0.0320	0.0875	0.839
+ GN^{Mamba}	**0.835**	**22.786**	**0.0314**	**0.0841**	**0.844**

Readers interested in additional ablations on specific components of the Mamba block (bidirectional SSM, Generalised-Hilbert scan), are directed to our paper on the original single-contrast GAMBAS model [15].

B Hyperparameter Tuning

L1 weighting was selected based on results from the following experiments (Table 3):

Table 3. Performance of GAMBAS+ with adversarial weight kept at 1, and L1 weights varying from 100, 200, 300, and 400.

λ L1	SSIM (↑)	PSNR (↑)	MAE (↓)	NMSE (↓)	Score (↑)
100	0.817	22.507	0.0327	0.0895	0.830
200	0.812	22.543	0.0327	0.0886	0.827
300	**0.824**	**22.605**	**0.0322**	**0.0875**	**0.836**
400	0.823	22.574	0.0323	0.0883	0.834

Weighting for the perceptual loss was chosen by considering previous literature, where this value is often a fraction of the adversarial loss [34]. SSIM weighting was selected by ensuring the magnitude of this loss is close, but below that of the L1 loss. This ensured strong contribution without overpowering the primary voxel-wise loss.

C Ensembling Experiments

To assess whether average outputs from multiple models yielded better results than an individual GAMBAS+ model, we performed the following experiments (Table 4):

Table 4. Performance of a single GAMBAS+ model compared an ensemble of 2, 3, 4, and 5 models. Top 5 rows: voxel-wise metrics for full volume outputs; bottom 5 rows: voxel-wise metrics for skull-stripped outputs

Image	# Models	SSIM (↑)	PSNR (↑)	MAE (↓)	NMSE (↓)	Score (↑)
Full volume	1	0.835	22.786	0.0314	0.0841	0.844
	2	0.838	22.984	0.0307	0.0804	0.847
	3	0.839	23.036	0.0306	0.0795	0.849
	4	0.840	23.086	0.0304	0.0787	0.849
	5	0.840	23.097	0.0304	0.0785	0.849
Skull- stripped	1	0.936	26.607	0.0147	0.0519	0.932
	2	0.938	26.266	0.0159	0.0580	0.931
	3	0.938	26.341	0.0158	0.0569	0.932
	4	0.939	26.172	0.0162	0.0593	0.931
	5	0.939	26.111	0.0164	0.0597	0.931

When examining full volumes, we observed a large performance increase going from one model to two models, and incremental increases from two models upwards. Importantly, to assess whether model averaging enhanced details within the brain itself, we recalculated all voxel-wise metrics following skull-stripping of model outputs using HD-BET [35]. In this scenario, we observed minimal benefit from model averaging, suggesting that improvements seen in full volumes were primarily confined to skull tissue. Nonetheless, given that skull-stripping was not part of the ULF-EnC challenge evaluation protocol, we opted to include model averaging. Ultimately, we chose to perform ensembling using four models, as this yielded the greatest improvement while keeping inference time within the requirements of the challenge.

References

1. Arnold, T.C., Freeman, C.W., Litt, B., Stein, J.M: Low-field MRI: clinical promise and challenges. J. Magn. Reson. Imaging **57**(1), 25–44 (2023). https://doi.org/10.1002/jmri.28408
2. Campbell-Washburn, A.E., Ramasawmy, R., Restivo, M.C., et al.: Opportunities in interventional and diagnostic imaging by using high-performance low-field-strength MRI. Radiology **293**(2), 384–393 (2019). https://doi.org/10.1148/radiol.2019190452
3. Dayarathna S., Islam Kh T., Uribe S., Yang G., Hayat M., Chen Z.: Deep learning based synthesis of MRI, CT and PET: review and analysis. Med. Image Anal. **92** (2024). https://doi.org/10.1016/j.media.2023.103046
4. Iglesias, J.E., et al.: Joint super-resolution and synthesis of 1 mm isotropic MP-RAGE volumes from clinical MRI exams with scans of different orientation, resolution and contrast. Neuroimage **237**, 118206 (2021). https://doi.org/10.1016/j.neuroimage.2021.118206
5. Iglesias, J.E., et al.: Quantitative brain morphometry of portable low-field-strength MRI using super-resolution machine learning. Radiology **306**(3), e220522 (2023). https://doi.org/10.1148/radiol.220522
6. Islam, K.T., et al.: Improving portable low-field MRI image quality through image-to-image translation using paired low- and high-field images. Sci. Rep. **13**(1) (2023). https://doi.org/10.1038/s41598-023-48438-1
7. Tapp, A., et al.: Super-field MRI Synthesis for Infant Brains Enhanced by Dual Channel Latent Diffusion. In: Linguraru, M.G., et al. (ed.) Proceedings of Medical Image Computing and Computer Assisted Intervention – MICCAI 2024, LNCS 15003, pp. 444–454. Springer, Cham (2024). https://doi.org/10.1007/978-3-031-72384-1_42
8. Dayarathna, S., Islam, K.T., Chen, Z.: Ultra low-field to high-field MRI translation using adversarial diffusion. In: 2024 IEEE International Symposium on Biomedical Imaging (2024)
9. Wang, Z., Wu, L.: Theoretical analysis of the inductive biases in deep convolutional networks. In: Thirty-Seventh Conference on Neural Information Processing Systems (2023)
10. Dosovitskiy, A., et al.: An image is worth 16 × 16 words: transformers for image recognition at scale. In: International Conference on Learning Representations (2021)
11. Dalmaz, O., Yurt, M., Çukur, T.: ResViT: residual vision transformers for multimodal medical image synthesis. IEEE Trans. Med. Imaging **41**(10), 2598–2614 (2022). https://doi.org/10.1109/TMI.2022.3167808
12. Gu, A., Dao, T.: Mamba: linear-time sequence modeling with selective state spaces. arXiv preprint arXiv:2312.00752 (2023)
13. Ma, J., Li, F., Wang, B.: U-Mamba: enhancing long-range dependency for biomedical image segmentation. arXiv preprint arXiv:2401.04722 (2024)
14. Heidari, M., et al.: Computation-efficient era: a comprehensive survey of state space models in medical image analysis. arXiv preprint arXiv:2406.03430 (2024)
15. Baljer, L., Briski, U., Leech R., Bourke, N., Donald, K., et al.: GAMBAS: generalised-Hilbert mamba for super-resolution of paediatric ultra-low-field MRI. In: Proceedings of Machine Learning Research, pp. 1–18 (2025)
16. Lucas, A., Arnold, T.C., Okar, S.V., et al.: Multisequence 3-T image synthesis from 64-mT low-field-strength MRI using generative adversarial networks in multiple sclerosis. Radiology **315**(1) (2025). https://doi.org/10.1148/radiol.233529

17. Dayarathna, S., et al.: McCaD: multi-contrast MRI conditioned, adaptive adversarial diffusion model for high-fidelity MRI synthesis. In: 2025 IEEE/CVF Winter Conference on Applications of Computer Vision (WACV), pp. 670–679. IEEE (2025)
18. Gu, A., Goel, K., Ré, C.: Efficiently modeling long sequences with structured state spaces. In: The International Conference on Learning Representations (ICLR) (2022)
19. Zhou, W., Kamata, S., Wang, H., Wong, M.S., Hou, H.C.: Mamba-in-Mamba: centralized mamba-cross-scan in tokenized Mamba model for hyperspectral image classification. Neurocomputing **613**, 128751 (2025). https://doi.org/10.1016/j.neucom.2024.128751
20. Warren, M.S., Salmon, J.K.: A parallel hashed Oct-Tree N-body algorithm. In: Proceedings of the 1993 ACM/IEEE Conference on Supercomputing, Portland, OR, USA, pp. 12–21 (1993)
21. Červený, J.: Gilbert. https://github.com/jakubcerveny/gilbert (2019). Accessed 01 Dec 2024
22. Zhu, J.-Y., Park, T., Isola, P., Efros, A.A.: Unpaired image-to-image translation using cycle-consistent adversarial networks. In: 2017 IEEE International Conference on Computer Vision (ICCV), pp. 2242–2251 (2017). https://doi.org/10.1109/ICCV.2017.244
23. Atli, O.F., Kabas, B., Arslan, F., Yurt, M., Dalmaz, O., Çukur, T.: I2I-Mamba: multi-modal medical image synthesis via selective state space modeling. arXiv:2405.14022 (2024)
24. Kumar, A., Fulham, M., Feng, D., Kim, J.: o-learning feature fusion maps from PET-CT images of lung cancer. IEEE Trans. Med. Imaging **39**(1), 204–217 (2020)
25. Isola, P., Zhu, J.-Y., Zhou, T., Efros, A.A.: Image-to-image translation with conditional adversarial networks. In: 2017 IEEE Conference on Computer Vision and Pattern Recognition (CVPR), pp. 5967–5976 (2017). https://doi.org/10.1109/CVPR.2017.632
26. Zhang, R., Isola, P., Efros, A.A., Shechtman, E., Wang, O.: The unreasonable effectiveness of deep features as a perceptual metric. In: Proceedings of the IEEE Conference on Computer Vision and Pattern Recognition (CVPR) (2018)
27. Mei, X., et al.: RadImageNet: an open radiologic deep learning research dataset for effective transfer learning. Radiol. Artif. Intell. **4**(5) (2022). https://doi.org/10.1148/ryai.210315
28. Cardoso, M.J., Li, W., Brown, R., Ma, N., Kerfoot, E., Wang, Y., et al.: MONAI: an open-source framework for deep learning in healthcare. arXiv preprint arXiv:2211.02701 (2022)
29. Ganaie, M.A., Hu, M., Malik, A.K., Tanveer, M., Suganthan, P.N.: Ensemble deep learning: a review. Eng. Appl. Artif. Intell. **115**, 105151 (2022). https://doi.org/10.1016/j.engappai.2022.105151
30. Wang, C., et al.: Ensembling diffusion models via adaptive feature aggregation. In: The Thirteenth International Conference on Learning Representations (2025)
31. Islam Kh T., Zhong, S., Peiris, H., Zakavi, P., Dayarathna, S., Chen, Z.: Enhancing ultra-low-field MRI with paired high- field MRI comparisons for brain imaging (ULF-EnC) (2025). https://doi.org/10.5281/zenodo.15259777
32. Jenkinson, M., Smith, S.: A global optimisation method for robust affine registration of brain images. Med. Image Anal. **5**(2), 143–156 (2001). https://doi.org/10.1016/S1361-8415(01)00036-6
33. King's College London: King's Computational Research, Engineering and Technology Environment (CREATE) (2022). https://doi.org/10.18742/rnvf-m076

34. Pinaya, W.H.L., et al.: Generative AI for medical imaging: extending the MONAI framework. arXiv preprint arXiv:2307.15208 (2023)
35. Isensee, F., et al.: Automated brain extraction of multi-sequence MRI using artificial neural networks. Hum. Brain Mapp. **2019**, 1–13 (2019). https://doi.org/10.1002/hbm.24750

Enhancing Ultra-Low-Field to High-Field MRI Using Multi-cycle GAN and Flow Matching Models

Jinghang Li, Bruno de Almeida, and Tamer S. Ibrahim(✉)

University of Pittsburgh, Pittsburgh, PA 15261, USA
tibrahim@pitt.edu

Abstract. There is an increasing interest in developing models capable of enhancing ultra-low-field (ULF) to high-field (HF) Magnetic Resonance Imaging (MRI) images, driven by the lower cost and greater portability of ULF scanners. In this work, we explore two distinct deep learning approaches to improve the visual quality of ULF MRI: a 3D adaptation of the multi-cycle Generative Adversarial Network (GAN), and a 2D flow matching model. Both methods incorporate autoregressive strategies, where each predicted slice is conditioned on previously generated slices to enforce spatial consistency and anatomical coherence across volumes. These models are trained and evaluated on paired datasets containing both ULF (64mT) and HF (3T) images from the *2025 - Ultra-Low-Field MRI Image Enhancement Challenge* (ULF-EnC). Our experiments indicate that both models can enhance MRI quality, with the flow matching approach yielding better results, demonstrating the potential to bridge the quality gap between accessible low-field acquisitions and clinically preferred high-field MRI, potentially improving diagnostic value in resource-limited settings.

Keywords: Flow Matching · Multicycle GAN · ultra-low-field (ULF) MRI · 64mT MRI · high-field (HF) MRI · 3T MRI · image enhancement · paired MRI data · deep learning · medical imaging · diagnostic imaging · brain imaging · healthcare accessibility · image quality improvement

1 Introduction

Magnetic Resonance Imaging (MRI) is an imaging technology with excellent capabilities for observing soft tissue contrast in a non-invasive manner. It is widely used to detect and monitor tumors, traumatic injuries, neurodegenerative diseases, vascular abnormalities, and musculoskeletal disorders [1–4]. Clinically, most MRI machines operate at high-field (HF) strength, such as 1.5 Tesla (T) and 3T [5]. However, while these scanners produce superior image quality, they

J. Li and B. de Almeida—These authors contributed equally to this work.

Z. Chen et al. (Eds.): ULF-EnC 2025, LNCS 16293, pp. 173–182, 2026.
https://doi.org/10.1007/978-3-032-23344-8_19

are costly to acquire and maintain, and are often unavailable in resource-limited settings. In contrast, ultra-low-field (ULF) MRI scanners offer a more affordable and portable alternative, although they produce images with significantly lower spatial resolution and signal-to-noise ratio (SNR) [6,7]. This quality gap presents a significant challenge for clinical adoption of ULF systems.

Recently, machine learning techniques have emerged as a promising solution to bridge this gap, enabling enhancement of medical images acquired at low field strengths without additional scanning burden on patients. In this work, we investigate two popular generative models: Generative Adversarial Network (GAN) [8] and flow matching model [9].

1.1 Generative Adversarial Networks for Medical Imaging

GAN has become popular for image enhancement due to its ability to generate high-quality, realistic images by learning complex data distributions. The framework consists of the simultaneous training of two networks: a generator that produces synthetic images, and a discriminator that distinguishes real from generated images. Building on this foundation, CycleGAN introduced bidirectional domain translation with cycle-consistency constraints, ensuring that translating an image to another domain and back reconstructs the original [10]. This property is particularly crucial in medical imaging, where preserving anatomical structures and diagnostic information is paramount.

The medical imaging community has successfully applied GAN-based approaches to various challenges, including artifact removal, cross-field strength translation (e.g., from 3T to 7T MRI), and cross-modality synthesis (e.g., MRI to Computed Tomography (CT)) [11–14]. These successes motivated our adoption of the multi-cycle GAN framework, originally proposed for MRI-to-CT translation [14], and adapted it to the ULF-to-HF enhancement problem with the preservation of 3D spatial context.

1.2 Flow Matching as an Alternative Paradigm

While GANs have proven effective, they can suffer from training instability and mode collapse. Denoising diffusion models emerged as a robust alternative, offering simpler optimization objectives and more stable training [15,16]. These models generate data by learning to reverse a forward process that gradually adds noise to real samples. However, their sampling process is computationally expensive, requiring many iterative steps [17–20].

Flow matching extends the diffusion framework to continuous-time deterministic dynamics, formulating generation as integrating an ordinary differential equation (ODE) that transports a simple prior distribution to the target data distribution [9]. By directly learning the velocity field that defines this transformation, flow matching inherits the stability benefits of diffusion models while enabling more efficient sampling, including the possibility of single-step generation [21]. This efficiency-stability trade-off makes flow matching particularly attractive for medical imaging applications where both quality and computational resources are critical considerations.

1.3 Our Contributions

In this work, we present a comprehensive comparison of two generative paradigms for ULF-to-HF MRI enhancement. Our key contributions include:

- A 3D adaptation of the multi-cycle GAN that processes volumetric slabs to capture spatial context across adjacent slices.
- Implementation of a 2D flow matching model optimized for medical image translation with efficient sampling strategies.
- Integration of autoregressive conditioning in both frameworks to enforce volumetric consistency.

2 Methods

2.1 Dataset and Preprocessing

We utilized paired 64 mT and 3T images from the *2025 Ultra-Low-Field MRI Image Enhancement Challenge* (ULF-EnC) [22–25], acquired from healthy participants using a Siemens Biograph mMR (3T) scanner and a Hyperfine Swoop (64 mT) scanner at Monash Biomedical Imaging, Monash University, Australia. Three contrasts were acquired for each participant in both domains: T1-weighted (T1w), T2-weighted (T2w), and Fluid-Attenuated Inversion Recovery (FLAIR).

The challenge provided 50 paired ULF-HF images for training and 10 ULF validation images for evaluation. All images underwent standardized preprocessing by the challenge organizers: initial quality control to exclude corrupted data, conversion from DICOM to NIfTI format, bias field correction using FSL-FAST [26], resampling to 1mm^3 isotropic resolution, and co-registration using FSL-FLIRT [27]. This preprocessing ensured consistent spatial alignment between paired volumes, facilitating supervised training of our models.

2.2 Multi-cycle GAN

We developed a 3D adaptation of the multi-cycle GAN previously used to synthesize CT images from MRI acquisitions [14]. In addition to the two generators typically found in the CycleGAN framework, which translate images between paired domains, this model introduces a third translation step that revisits the synthetic images. Consequently, in addition to the Real-Cycle Consistency Loss ($\mathcal{L}_{RCC}$), which enforces consistency between the original input and its reconstruction after one forward–backward cycle, a Pseudo-Cycle Consistency Loss ($\mathcal{L}_{PCC}$) was introduced to encourage consistency between the first-pass synthetic image and its counterpart after this extra pseudo cycle.

This multi-cycle approach aims to improve training stability and preserve spatial continuity across slices. In addition, a Domain Control Loss ($\mathcal{L}_{DC}$) is a

L_1 identity loss applied to images that are already in the target domain (and to their pseudo-cycle outputs), preventing the generators from needlessly modifying in-domain inputs. Adversarial supervision is split into the standard generator and discriminator losses, Generator Adversarial Loss ($\mathcal{L}_{adv\_G}$) and Discriminator Adversarial Loss ($\mathcal{L}_{adv\_D}$), plus the Domain Control Adversarial Loss ($\mathcal{L}_{adv\_DC}$), which is applied to the identity outputs to ensure realism without corrupting domain-specific features. In summary, the total generator and discriminator loss equations are defined as:

$$\mathcal{L}_G = \alpha\mathcal{L}_{RCC} + \beta\mathcal{L}_{PCC} + \delta\mathcal{L}_{DC} + \gamma\mathcal{L}_{adv\_G} + \epsilon\mathcal{L}_{adv\_DC} \tag{1}$$

$$\mathcal{L}_D = \mathcal{L}_{adv\_D} \tag{2}$$

Instead of predicting 2D slices, our generators produce slabs of four contiguous slices. Initially, images were normalized across the whole volume, using only voxels within the 1st and 99th percentiles (excluding voxels with intensity zero) to mitigate the influence of outliers. Next, we generated all possible slab combinations with overlap. The generator and discriminator architectures closely follow those described in [14]: the generators are based on a U-Net with seven layers, where the fourth layer incorporates a Z-Net structure. In this work, we investigated the influence of the number of channels per layer: a 32-channel model uses 32 channels in the first layer, while a 64-channel model starts with 64 channels. In the encoder, the number of channels is doubled at each downsampling stage, whereas in the decoder, the number of channels is halved at each upsampling stage.

Our model can be configured to predict a single contrast modality from multiple input contrasts. In such cases, an additional multilayer perceptron is employed to reconstruct the remaining modalities from the predicted output. When the number of predicted contrasts matches the number of input contrasts, this mapping is replaced by an identity operation.

We also explored an autoregressive approach: starting from the previously trained model, we introduced an additional training stage in which a single slice is predicted conditioned on previously generated slices. To make predictions more robust, the initial training steps used the ground-truth slices in place of previous predictions, and the probability of using ground-truth data was gradually decreased as training progressed. Furthermore, in approximately 30% of cases, random Gaussian noise was added to the previous slices to simulate prediction errors and improve model robustness, while in another 30% of cases the previous slices were entirely replaced by zeros to force the network to recover from missing contextual information. Taking advantage of the absence of relevant anatomical structures at the top of the brain volume, training was performed in a top-to-bottom slice order.

Training Strategy: We adopted $\alpha = 30$, $\beta = 30$, $\delta = 10$, $\gamma = 1$, and $\epsilon = 10$ to balance the contributions of the individual loss terms during training. For the baseline cases with 32- and 64-channel configurations, the model was trained for 100 epochs. During the first 20 epochs, only the $\mathcal{L}_{RCC}$ and $\mathcal{L}_{adv\_G}$ losses were applied; in the following 20 epochs, the $\mathcal{L}_{PCC}$ loss was additionally included; and in the remaining epochs, all loss terms were jointly optimized. In these baseline settings, only axial slices per slab were used, and each contrast was predicted using an independent generator, with all available contrasts serving as inputs.

Three additional experimental cases were explored for each configuration, always starting from the baseline weights and using similar hyperparameters and loss scheduling: **32ch/AR** and **64ch/AR**: autoregressive approach; **32ch/MV** and **64ch/MV**: multi-view approach, where an additional training round used slabs in axial, sagittal, and coronal orientations. **32ch/2St** and **64ch/2St**: two-stage approach, where a second generator, receiving all contrasts predicted by baseline generators as input, was trained to further enhance the outputs of the initial generators.

2.3 Flow Matching

Our flow matching approach reformulates the ULF-to-HF translation as learning a continuous transformation through an ODE. This framework provides simpler training objective than GANs while enabling efficient sampling strategies.

Model Architecture. We employed a U-Net architecture with three feature levels (256, 256, and 512 channels), incorporating self-attention mechanisms at the deepest level for capturing long-range dependencies. The network processes 3-channel inputs: the current noisy sample x_t, the ULF conditioning slice x_c, and the previous frame of the HF image x_p. This design enables the model to maintain both local in-plane image quality and global volumetric consistency.

Training Objective. Following the flow matching framework, we define a linear interpolation path between noise $x_0 \sim \mathcal{N}(0, \mathbf{I})$ and target image x_1:

$$x_t = (1-t)x_0 + tx_1, \quad t \in [0,1] \tag{3}$$

The velocity field for this path is $v_t = x_1 - x_0$. Our model v_θ is trained to predict this velocity given the intermediate state and conditioning:

$$\mathcal{L} = \mathbb{E}\left[|v_\theta(x_t, x_c, x_p) - (x_1 - x_0)|_2^2\right] \tag{4}$$

This objective provides a direct supervision signal at every timestep, leading to stable and efficient training.

Sampling Strategies. During inference, we generated HF images by integrating the learned velocity field:

$$x_1 = x_0 + \int_0^1 v_\theta(x_t, x_c, x_p)\, dt \tag{5}$$

We implemented both single-step and multi-step integration using Euler's method. The multi-step approach provides better quality at the cost of increased computation, while single-step generation offers real-time performance. Similar to the multi-cycle GAN, we incorporated autoregressive conditioning to maintain volumetric consistency across slices.

2.4 Evaluation Framework

Model performance was assessed using four complementary metrics: Structural Similarity Index Measure (SSIM) for structural similarity, Peak Signal-to-Noise Ratio (PSNR) for signal fidelity, Mean Absolute Error (MAE) for absolute error, and Normalized Mean Squared Error (NMSE) for normalized error. These metrics were computed both globally and within brain-skull masked regions. The challenge's final score combines these metrics:

$$\text{Final} = 0.7 \times \text{SSIM} + 0.1 \times \text{PSNR}/32 + 0.1 \times (1 - \text{MAE}) + 0.1 \times (1 - \text{NMSE}) \quad (6)$$

This weighted combination emphasizes structural preservation while considering multiple aspects of image quality, providing a comprehensive assessment aligned with clinical relevance.

3 Results

Table 1 presents the comprehensive evaluation results across all model variants. The flow matching approach with autoregressive conditioning and 5-step ODE integration achieved the best overall performance, with a final score of 0.780 globally and 0.645 in masked regions. This represents a significant improvement over the best multi-cycle GAN configuration (64-channel baseline: 0.768 global, 0.628 masked). Figures 1, 2, and 3 showcase axial, coronal, and sagittal enhancements from a 64 mT case across the three modalities (FLAIR, T1w, and T2w) using the flow matching model for the cases ODE (5 steps) and ODE AR (5 steps).

The scores in Table 1 were provided by the challenge organizers after submitting our predicted validation images through their evaluation website. For each model, the reported scores correspond to the average performance across the T1w, T2w, and FLAIR contrasts.

4 Discussion

Our results demonstrated that both multi-cycle GAN and flow matching could effectively enhance ULF MRI images.

The multi-cycle GAN showed consistent but limited improvement with increased model complexity. The marginal gains from doubling channel capacity (32 to 64) suggested that the bottleneck lay in the training dynamics rather than

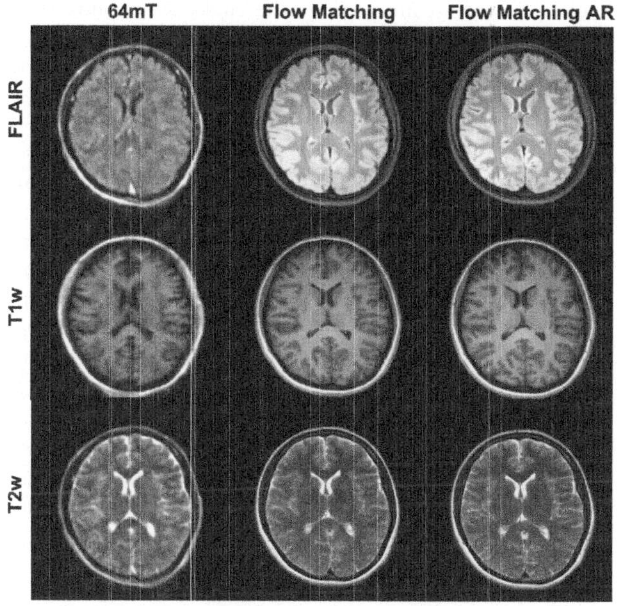

Fig. 1. Axial view comparison of flow matching enhancement results with and without autoregressive conditioning (ODE and ODE AR with 5 steps).

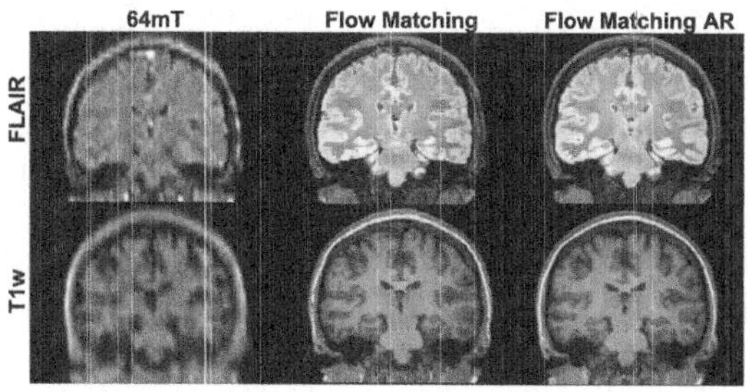

Fig. 2. Coronal view comparison of flow matching enhancement results with and without autoregressive conditioning (ODE and ODE AR with 5 steps).

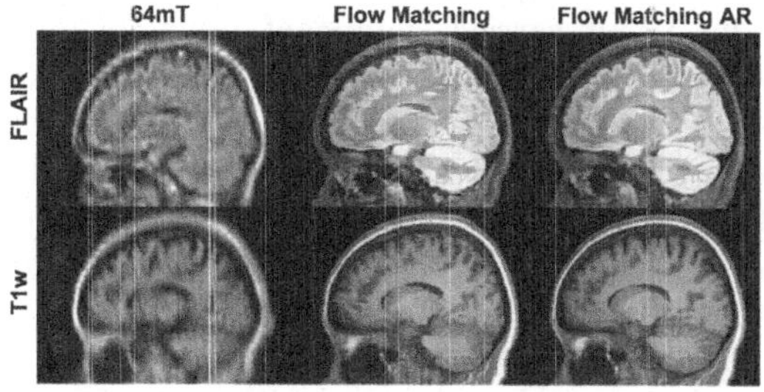

Fig. 3. Sagittal view comparison of flow matching enhancement results with and without autoregressive conditioning (ODE and ODE AR with 5 steps).

Table 1. Enhancement scores computed by the challenge organizers for each model, reported separately for global and masked (brain and skull) regions. The last case *Testing Phase* represents the challenge scores for our submitted model (only the masked case was provided).

Model	Case	Global					Masked				
		SSIM	PSNR	MAE	NMSE	Final	SSIM	PSNR	MAE	NMSE	Final
Multi-cycle GAN	32-ch	0.739	19.942	0.046	0.138	0.761	0.534	26.232	0.101	0.316	0.614
	32-ch/AR	0.662	17.644	0.064	0.226	0.690	0.398	23.022	0.141	0.519	0.485
	32-ch/MV	0.720	19.521	0.049	0.153	0.745	0.506	25.613	0.107	0.351	0.588
	32-ch/2St	0.731	19.819	0.047	0.142	0.755	0.535	25.942	0.103	0.324	0.613
	64-ch	0.747	20.168	0.045	0.132	0.768	0.550	26.458	0.098	0.302	0.628
	64-ch/AR	0.670	18.408	0.056	0.193	0.702	0.425	24.249	0.122	0.422	0.519
	64-ch/MV	0.724	19.942	0.047	0.14	0.750	0.530	25.934	0.102	0.32	0.610
	64-ch/2St	0.743	19.956	0.046	0.139	0.764	0.549	26.247	0.101	0.318	0.624
Flow Matching	ODE (5 steps)	0.641	17.478	0.065	0.266	0.670	0.341	24.132	0.144	0.632	0.437
	ODE AR (1 step)	0.746	20.520	0.044	0.121	0.770	0.561	26.939	0.096	0.282	0.639
	ODE AR (5 steps)	**0.760**	**20.594**	**0.043**	**0.119**	**0.780**	**0.566**	**27.359**	**0.094**	**0.275**	**0.645**
Testing Phase	ODE AR (5 steps)	–	–	–	–	–	**0.545**	**26.452**	**0.101**	**0.133**	**0.641**

model expressiveness. The failure of autoregressive conditioning to improve performance indicated that the adversarial training struggled to effectively leverage temporal dependencies. This could have been due to the increased complexity of the optimization landscape when combining cycle-consistency constraints with autoregressive conditioning.

When processing slices independently, the flow matching model produced high-quality in-plane enhancements but suffered from poor anatomical continuity across slices—a critical limitation for volumetric medical images. However, the introduction of autoregressive conditioning dramatically resolved this issue, improving both quantitative metrics (SSIM: 0.641 → 0.760) and visual coherence across the volume.

Our most notable finding was the near-linearity of the learned transformation trajectory from Gaussian noise to HF images when conditioned on ULF input. Remarkably, the model was able to produce images of reasonable quality in a single integration step, achieving a final score of 2.758 compared to 2.775 with five steps. This result carried significant practical implications: it indicated that the ULF-to-HF mapping—despite its apparent complexity—could be well-approximated by a relatively simple transformation, enabling near real-time image enhancement during image acquisition. The conditioning ULF image exerted sufficiently strong guidance that the velocity field remained nearly constant along the integration path, allowing direct transitions from noise to target with only minimal quality degradation.

Acknowledgments. We gratefully acknowledge the team from Monash University for providing the training data and hosting the challenge. This research was supported in part by the University of Pittsburgh Center for Research Computing and Data,

RRID:SCR_022735, through the resources provided. Specifically, this work used the H2P cluster, which is supported by NSF award number OAC-2117681.

References

1. Ogawa, S., Lee, T.-M., Nayak, A.S., Glynn, P.: Oxygenation-sensitive contrast in magnetic resonance image of rodent brain at high magnetic fields. Magn. Reson. Med. **14**(1), 68–78 (1990)
2. Wen, P.Y., et al.: Updated response assessment criteria for high-grade gliomas: response assessment in neuro-oncology working group. J. Clin. Oncol. **28**(11), 1963–1972 (2010)
3. Jack, C.R., Jr., et al.: The Alzheimer's disease neuroimaging initiative (ADNI): MRI methods. J. Magn. Reson. Imaging **27**(4), 685–691 (2008)
4. Deyle, G.D.: The role of MRI in musculoskeletal practice: a clinical perspective. J. Man. Manip. Ther. **19**(3), 152–161 (2011)
5. Pai, A., Shetty, R., Hodis, B., Chowdhury, Y.S.: Magnetic resonance imaging physics. In: StatPearls. StatPearls Publishing (2023)
6. Arnold, T.C., Freeman, C.W., Litt, B., Stein, J.M.: Low-field MRI: clinical promise and challenges. J. Magn. Reson. Imaging **57**(1), 25–44 (2023)
7. Altaf, A., et al.: Applications, limitations and advancements of ultra-low-field magnetic resonance imaging: a scoping review. Surg. Neurol. Int. **15**, 218 (2024)
8. Goodfellow, I.J., et al.: Generative adversarial nets. In: Advances in Neural Information Processing Systems, vol. 27 (2014)
9. Lipman, Y., Chen, R.T.Q., Ben-Hamu, H., Nickel, M., and Le, M.: Flow matching for generative modeling. arXiv preprint arXiv:2210.02747 (2022)
10. Zhu, J.-Y., Park, T., Isola, P., Efros, A.A.: Unpaired image-to-image translation using cycle-consistent adversarial networks. In: Proceedings of the IEEE International Conference on Computer Vision, pp. 2223–2232 (2017)
11. Diniz, E., Santini, T., Karim, H., Aizenstein, H.J., Ibrahim, T.S.: Cross-modality image translation of 3 Tesla magnetic resonance imaging to 7 Tesla using generative adversarial networks. Hum. Brain Mapp. **46**(9), e70246 (2025)
12. Nie, D., et al.: Medical image synthesis with deep convolutional adversarial networks. IEEE Trans. Biomed. Eng. **65**(12), 2720–2730 (2018)
13. Wolterink, J.M., Dinkla, A.M., Savenije, M.H.F., Seevinck, P.R., van den Berg, C.A.T., Išgum, I.: Deep MR to CT synthesis using unpaired data. In: Simulation and Synthesis in Medical Imaging. LNCS, vol. 10557, pp. 14–23. Springer (2017). https://doi.org/10.1007/978-3-319-68127-6_2
14. Liu, Y., et al.: CT synthesis from MRI using multi-cycle GAN for head-and-neck radiation therapy. Comput. Med. Imaging Graph. **91**, 101953 (2021)
15. Song, Y., Sohl-Dickstein, J., Kingma, D.P., Kumar, A., Ermon, S., Poole, B.: Score-based generative modeling through stochastic differential equations. arXiv preprint arXiv:2011.13456 (2020)
16. Ho, J., Jain, A., Abbeel, P.: Denoising diffusion probabilistic models. In: Advances in Neural Information Processing Systems, vol. 33, pp. 6840–6851 (2020)
17. Song, J., Meng, C., Ermon, S.: Denoising diffusion implicit models. arXiv preprint arXiv:2010.02502 (2020)
18. Ma, X., Fang, G., Wang, X.: Deepcache: accelerating diffusion models for free. In: Proceedings of the IEEE/CVF Conference on Computer Vision and Pattern Recognition, pp. 15762–15772 (2024)

19. Lu, C., Zhou, Y., Bao, F., Chen, J., Li, C., Zhu, J.: DPM-solver: a fast ODE solver for diffusion probabilistic model sampling in around 10 steps. In: Advances in Neural Information Processing Systems, vol. 35, pp. 5775–5787 (2022)
20. Liu, L., Ren, Y., Lin, Z., Zhao, Z.: Pseudo numerical methods for diffusion models on manifolds. arXiv preprint arXiv:2202.09778 (2022)
21. Kornilov, N., Mokrov, P., Gasnikov, A., Korotin, A.: Optimal flow matching: learning straight trajectories in just one step. In: Advances in Neural Information Processing Systems, vol. 37, pp. 104180–104204 (2024)
22. Islam, K.T., et al.: Improving portable low-field MRI image quality through image-to-image translation using paired low- and high-field images. Sci. Rep. **13**(1), 21183 (2023). https://doi.org/10.1038/s41598-023-48438-1
23. Dayarathna, S., et al.: Deep learning based synthesis of MRI, CT and PET: review and analysis. Med. Image Anal. **92**, 103046 (2024). https://doi.org/10.1016/j.media.2023.103046
24. Dayarathna, S., et al.: McCaD: multi-contrast MRI conditioned, adaptive adversarial diffusion model for high-fidelity MRI synthesis. In: 2025 IEEE/CVF Winter Conference on Applications of Computer Vision (WACV), pp. 670–679. IEEE (2025). https://doi.org/10.1109/WACV61041.2025.00075
25. Dayarathna, S., et al.: Ultra low-field to high-field MRI translation using adversarial diffusion. In: 2024 IEEE International Symposium on Biomedical Imaging (ISBI), pp. 1–4. IEEE (2024). https://doi.org/10.1109/ISBI56570.2024.10635808
26. Zhang, Y., Brady, M., Smith, S.: Segmentation of brain MR images through a hidden Markov random field model and the expectation-maximization algorithm. IEEE Trans. Med. Imaging **20**(1), 45–57 (2002)
27. Jenkinson, M., Bannister, P., Brady, M., Smith, S.: Improved optimization for the robust and accurate linear registration and motion correction of brain images. Neuroimage **17**(2), 825–841 (2002)

Author Index

Z. Chen et al. (Eds.): ULF-EnC 2025, LNCS 16293, pp. 183–184, 2026.
https://doi.org/10.1007/978-3-032-23344-8

The manufacturer's authorised representative in the EU is Springer
Nature Customer Service Centre GmbH, Europaplatz 3, 69115 Heidelberg,
Germany. If you have any concerns regarding our products, please
contact ProductSafety@springernature.com

Printed and bound by CPI Group (UK) Ltd, Croydon, CR0 4YY
03/08/2026
02184503-0001